Breeding, Rearing and Feeding Cheviot and Black Faced Sheep

by a Lammermuir Farmer

with an introduction by Jackson Chambers

This work contains material that was originally published in 1827.

This publication is within the Public Domain.

This edition is reprinted for educational purposes
and in accordance with all applicable Federal Laws.

Introduction Copyright 2018 by Jackson Chambers

Self Reliance Books

Get more historic titles on animal and stock breeding, gardening and old fashioned skills by visiting us at:

Introduction

I am pleased to present yet another practical title on breeding and raising livestock.

The work is in the Public Domain and is re-printed here in accordance with Federal Laws.

As with all reprinted books of this age that are intended to perfectly reproduce the original edition, considerable pains and effort had to be undertaken to correct fading and sometimes outright damage to existing proofs of this title. At times, this task is quite monumental, requiring an almost total "rebuilding" of some pages from digital proofs of multiple copies. Despite this, imperfections still sometimes exist in the final proof and may detract from the visual appearance of the text.

I hope you enjoy reading this book as much as I enjoyed making it available to readers again.

Jackson Chambers

PREFACE.

———

$\mathfrak{A}$T a time when knowledge of every kind is so widely diffused, and all useful information communicated through the most direct channels to the most remote parts of the nation, it may be wondered that one so obscurely situated as I am, should attempt to bring myself into the notice of the world. Publications upon every subject in which the common benefit of mankind is at all concerned are of the most frequent occurrence in the present enlightened age, when every thing which it is of importance for us to know, is treated with the strictest accuracy. But this is matter of no easy attainment, and accordingly many works, which in a darker age would have attracted public attention, are either little inquired after, or after a transient glance are consigned to oblivion. If the fate of the remarks which I have ventured to present is to be determined by the scrutiny of critical acuteness, I am afraid they too, unable to stand so severe a test, will sink under the load, and be condemned to unheard of obscurity. As they are not prepared for such a test, however, I hope they will not be tried by it, and will be considered only with respect to the utility which they may be calculated to afford to that class of people, to whom alone, if to any they can be found of any practical value. And if I have at all succeeded in accomplishing that end, I will consider myself as having gained a compensation for the trouble which I have been led to bestow, and an equivalent for all subordinate defects.

Important as are the additions which are successively made to our stores of practical, as well as speculative knowledge, it will be readily owned, by many at least, that our acquaintance has been too limited with several of the topics which come under our present notice. This will at once be granted by those who are the most competent judges, and whose opinion has received the support of their own fatal experience. The consequences of the prevalence of misguided judgment in conducting highly situated farms, and our real want of information concerning one of the most destructive diseases that ever spread desolation amongst the inferior creation, have, of late years, been marked in characters of blood. ' If I have in the least contributed to the establishment of more perfect guides to store farmers, and done any thing to banish that malady, which has so repeatedly diffused carnage among our flocks, I shall account myself highly fortunate. But if by some, and perhaps by all, it may be thought that I have not been sufficiently careful in advancing opinions, which a fuller investigation would not have allowed me to countenance, I would entreat their clemency on account of the disadvantages under which I have had to labour. The tract in which I have travelled has been hitherto but little trodden, and those few that have journeyed along it, have generally combated their way by the weapons of controversy. Occasional slips, which are to be found in works of almost every stamp, and which are no where

more likely to be found than in the remarks which I have submitted, will not
then I trust be considered in so very
unfavourable a light.

To avoid, however, as much as possible the imputation of any material
mistake, I have attested the truth of
the more important observations by the
evidence of unquestionable facts, and have
erected nothing upon unfounded specula-
tion. It may be, indeed, that in endeavouring to shun error from this source,
I have run into the opposite mistake, and
have become tedious by the detail of many
facts. But facts when fairly adduced are
" stubborn proofs," and substantiate con-
clusions which without them might be
looked upon as altogether hypothetical.
So that if in this respect I have laid
myself open to censure, the fault fortunately lies upon the safer side.

PREFACE.

I may take occasion to remark that the generality of my observations are entirely confined to those two breeds of sheep which are specified in the title-page. About these I have been conversant during the whole of my life, and consequently to them my attention has been more immediately directed. With respect to the manner, however, of accomplishing the cure of that pernicious disease of which it is my design particularly to treat, the method recommended is equally applicable to the other species that graze in more fertile districts. In this, and perhaps in other circumstances, I may have, in some measure accidentally, accomodated myself to those breeds, of which it was not my object professedly to treat.

I am aware that a part of my observations will be considered as having been anticipated by the recent publication of the honourable Captain Napier, and may consequently look upon that part as superfluous. My thoughts were, however, all arranged, and the plan that I was to pursue finally agreed to in my own mind before this publication had reached my hands; and therefore, to have omitted what had come under his notice, would have made the observations which I have to offer appear unfinished and unconnected. Though, indeed, the measures, in the recommendation of which I have had the honour to coincide with that author, are illustrated in a far superior manner to what I could have made any pretensions.

And in the instances in which I have taken it upon me to espouse a contrary conviction from that to which this honourable author has yielded his assent,

and wherein I have had occasion to dis-
agree from the opinions that Mr Hogg
has advanced on the diseases of sheep,
my conduct may perhaps be accused as
presumptive. It is, indeed, with feelings
of no small regret that I have perused
the passages in the works of these celebrated gentlemen which contain sentiments contradictory to those which have
received the sanction of my approval;
and it is with reluctance that I submit
to the judgment of others the statement
of my dissonance. But as the subjects
on which we differ necessarily come under my view, in my present plan, I have
thought it incumbent upon me, notwithstanding the weight of such high authority, to adhere to what my observation
has invariably determined me. For it is
only in those things wherein my actual
experience has confirmed different senti-

ments that I have given a decided dissent from their opinions. And whether
of the two. is supported by the clearest
evidence, and bears along with it the
most unequivocal marks of truth, it is
now for others to determine.

CONTENTS.

——

PRELIMINARY OBSERVATIONS.

As the following treatise is confined chiefly to Cheviot and Black-faced Sheep, or such as are bred and reared in high and upland districts; and as from the situation of the author, the remarks contained in it are more immediately applicable to these breeds in their relation to Lammermuir, it may not be improper, for the information of people remotely situated, to give a very brief account of that district or tract of country. Though it may appear uninteresting to those who are themselves inhabitants of this uninviting region, or who are intimately acquainted with its various parts, there are yet many whose distance precludes them from possessing the proper sources of information concerning it, as a district appropriated to sheep.

The county of Berwick naturally divides itself into two great divisions, which may be termed the high and the low parts of the county. The lower part (generally called the merse) comprehends all that fine and highly cultivated tract of land from the banks of the Tweed to where Lammermuir begins. This division must necessarily be very arbitrary as great incroachments have been made by the plough on lands which, not many years ago, were covered with heath, and which, for their immediate effects at least, had far better, in many places, been lying to the present moment in all the waste and untaught rudeness of nature.

Lammermuir, or the high part of the county, may be reckoned from St Abb's Head on the east, to Crookston burn at the foot of Clints-hill on the west; a distance of more than 30 miles. The western part of Lammermuir is much broader than the eastern, and justly admits of being compared to a cone, the base of which resting on Soutra hill and Lauderdale, thence stretches its irregular form towards the German ocean, till it is cut short at Coldingham and St Abb's Head.

Of the 285,440 English acres which the whole county comprehends, 175,734 are included in Lammermuir. This, though the most extensive, is by much the least valuable part, and has been subdivided into high and low Lammermuir. The latter forms a kind of middle district between the Merse, and the highest hills in Lammermuir, and which both from the favourable nature of pasturage and climate is in general fully qualified to support the Cheviot breed of sheep. A good proportion of the land has, at one period or another, been brought into a state of tillage, and the pasture hills are here and there interspersed with patches of benty grass, though heath is the most prevalent production.

The range of hills that compose the greatest and most upland part of Lammermuir are altogether covered with heath, but intersected by numerous streams of water, which occasionally with their tributary rivulets form small but rather beautiful glens, which are fertile enough to yield grass pastures. The wildness of this tract of country, the general barrenness of the soil, and the continued inclemency of the winter season, forbid any other breed than the Black-faced from participating of the produce of the unprolific land. In some places, indeed, where with the heath there is an intermixture of grass, the Cheviot have been reared with no inconsiderable success; but the other more hardy breed are better fitted for the barrenness and exposure of the mountains, can be kept with much less expense, and in general are ultimately productive of greater advantage in this high region.

There are few situations in Scotland where the soil is more sterile and the climate more rigidly severe, than in the range of the highest mountains in Lammermuir. Whatever improvements are adopted, and whatever sheep graze on them, will be attended with equal success in almost any part of this island. Though they are not so highly elevated as many of those in the Highlands of Scotland, and the cold consequently not so intense, yet here there are none, at least there are comparatively none, of those fertile vales which add so much to the beauty of these northern parts, into which the flocks are brought at the commencement of winter, and in which, sheltered from the storm by the surrounding mountains, they lie in all imaginable security. The hills of Lammermuir, on the contrary, are only occasionally intersected by a deep hollow, in which, on account of its narrowness, it would hardly be exempt from danger to put sheep in a violent storm, for fear of being overwhelmed in wreaths of snow, and in which even in mild weather they would sustain injury from the confined nature of its limits. The stock in the wild places of Lammermuir is, therefore, exposed on the naked mountains to all the severity which the season may inflict, (unless when huddled together in these narrow glens on the appearance of a storm,) and to withstand which the most hardy constitution is fully requisite.

The district of Lammermuir is altogether compounded of hills; some of which rise to no small elevation. Cribblaw raises its head more than 1600 feet above the level of the sea, and Clintshill about 1544. These, however, and all others in this ridge, are overtopped by Lammerlaw, which at the most elevated part is more than 1716 feet above the level of the sea.

It would appear that at some earlier period, there were few or no continual residing inhabitants in Lammermuir, and that the lower and more fertile parts of the county were previously occupied. The possession of these occupiers would gradually extend towards the more barren district, as they increased in numbers. And as at first property would be very limited, they would have no anxiety, neither perhaps would they dare, to live throughout all the inclemencies of the season, amidst the cold, rugged, and unprolific hills of Lammermuir. It is exceedingly likely that they would then continue to dwell in the more fertile district, and would annually on the return of spring remove their flocks to graze on the hills, which to them would then appear habitable. This, the names of many places, which would be erected as summer residences only in a temporary manner, seem clearly to indicate; as they are still denominated by the significant name of *shiel*, and would in former years be called *shieling-houses.*

From the accounts that have been transmitted of that early period, it appears, that so late as the fourteenth century, Lammermuir had been but very partially inhabited and that then a great part of it was occupied by deer and wild cattle. These it is probable would be gradually expelled from the frontiers of the hills, as they became pasturage for sheep, and would be driven to the wilder and more remote places. It does not appear, however, that, in the fourteenth century, the ground appropriated for sheep extended beyond the very skirts of Lammermuir; but as much as was appropriated was sufficiently occupied, and that the shepherds made no despicable appearance seems probable from the following passage of Redpath's Border History:

PRELIMINARY OBSERVATIONS.

" In the year 1372 lord Percy, the English warden, to revenge some losses and
insults, entered Scotland at the head of
7000 men, and having crossed the low
country of the Merse through one of its
most fertile spots, encamped at Dunse.
But his farther progress was stopt by
a contrivance of the shepherds and pea-
sants in that neighbourhood; who bethought themselves of employing in defence of their country, a very simple
sort of machine, which they commonly
made use of to frighten away from their
corn the deer and wild cattle that then
abounded in Lammermuir. These were
a kind of rattles made of pieces of dried
skin, distended around ribs of wood, that
were bended into a semicircular form,
and fixed to the end of long poles.
The bags being furnished with a few
hard pebbles, and vigorously shaken by
a rapid motion given to the poles, made
a hideous noise: and an unusual number of them being thus employed on the

tops of the adjacent hills, the horses of the English took fright; and breaking away from their keepers, ran wildly up and down the neighbouring fields, where they became a prey to the people of the country. The army also, awakened with the strange noise, and finding themselves in the morning deprived not only of their war horses, but also of many of their beasts of burden, retired on foot towards the Tweed in precipitation, and disorder, having left their baggage behind them."

It would be uninteresting, and I trust it will be unnecessary, to pursue our observations farther upon this district. We have seen that in the earlier ages, the most extensive range of it was accounted so extremely wild and altogether so unfavourable, as to be a place unfit both for the habitation of man, and for pasturage to sheep. The improvements of more civilized times, however, have, through the course of time, rendered it in a great measure adequate to both; and unproductive and unsheltered as the hills still are, even the most unpromising of them, are capable of keeping and of rearing Black-faced stock. The Cheviot can only with propriety be kept in the lower part of Lammermuir, or where the pasture partakes a good deal of grass; but may with equal propriety be kept in many places where the hills rise to a higher altitude than those in any part of Lammermuir, but where there is the inestimable advantage of extensive prolific glens.

It may now be a subject, of pleasing, perhaps of instructive research, to extend our observations for a little to the original of sheep. This has at least some connexion with the following treatise, as being a treatise upon sheep, and a certain nourishing part of the food of many of these original sheep, will be found to have immediate reference to that cure which it is our intention to prescribe for the

Rot. That I may not incur the imputation of plagiarism, however, I may here candidly acknowledge that the observations now to be offered, are chiefly derived from the information on that subject communicated by the celebrated Dr Pallas, professor of natural history in the Imperial academy of St Petersburgh.

—. The *ovis fera* or wild sheep, according to this author, is the parent of all our domestic varieties of sheep, however changed by servitude, climate, food, &c. in the hands of man; and this sheep he found in all its native vigour, boldness, and activity, inhabiting the vast chain of mountains, which run through the centre of Asia to the eastern sea, and the branches which it sends off to Great Tartary, China, and the Indies. This animal is denominated by the Siberians, *argali*, meaning wild sheep; and by the Russians *kamennoi barann*, or sheep of the rocks, from its ordinary place of abode.

It delights in the rocky mountains of the Asiatic chain above-mentioned, where it is ever to be met with basking in the rays of the sun; but it avoids the woods of the mountains, and every other object that would intercept the influence of the great luminary. Its food is the Alpine plants and shrubs, which it finds amongst the rocks.

The *argali* generally prefers a temperate climate, although he is to be found in Asiatic Siberia, as it is there furnished with its favourite bare rocks, sun-shine, and Alpine plants. It even makes its habitation in the cold eastern extremity of Siberia and Kamtschatka, which evidently demonstrates that nature has given a most extensive range to sheep in a wild state, equal to what has been allowed to the intelligent creation; a fact which shows that the sheep is confined to no certain latitude. The argali is so extremely wild that it gradually abandons a country as it becomes peopled.

PRELIMINARY OBSERVATIONS.

The ewe of the argali brings forth
before the melting of the snow. Her
young resembles much a young kid; save
only that in place of horns, they have a
large fat protuberance, and that they are
covered with a woolly hair, frizzled, and
of a dark grey. Notwithstanding that
the adult argali is wild and untameable,
the lamb may with little difficulty be
tamed when taken young, and brought
up like a domestic sheep.

The height of the argali is about that
of a small hart, but more robust and
nervous. Its form is less elegant than
that of the deer, and its legs and neck
shorter. Its head is much like that of
a ram, but its ears shorter. Its horns
are very large, and weigh in an adult
16*lbs.* The summer coat consists of
short hair, sleek, and resembling that of a
deer. The winter coat consists of wool
like down, mixed with hair, an inch and
a-half long, concealing at its roots a

fine woolly down, generally of a white colour.

Dr Pallas considers all the sheep that abound in Siberia, and the pastoral nations of Tartary, as belonging to the argali or wild sheep, and subdivided into four varieties. These are the long tailed, the short tailed, the fat tailed, and a mixed breed with longish tails, fat at the base, with a species of lean bony appendage tapering to a point. The fat tailed is the most abundant and the largest breed of sheep in the world. It is reared throughout all the temperate regions of Asia, from the frontiers of Europe to those of China. All the Normade hordes of Asia, the Turcomans, Kirguise, Calmouks, and Mongul Tartars rear it. The Persians and Hottentots also rear it in abundance. It exists in the purest and most unmixed state in the vast deserts of Great Tartary.

The flocks, therefore of all the Tartar hordes resemble each other by a large yellowish muzzle; by long hanging ears; by the large, spiral, wrinkled, angular and bent horns of the adult ram. A solid mass of fat is formed on the rump, and falling down, supplies the place of a tail, which being divided into two hemispheres, takes the form of the hips, with a little button of a tail in the middle, to be felt by the finger. This fat protuberance amounts to from 20*lbs.* to 40*lbs.*

The southern Tartar flocks enjoy a moderate winter with regard to cold, though they pass it in the open air, living mostly on dry stalks, especially those of the half dry wormwood, which is abundant in the more elevated situations. There is likewise every where found an efflorescence of nitron with sea salt.

They are conducted by their masters in the spring to pastures, rich in rising plants and flowers; and are brought into a most palatable and favourite pasturage, sprinkled with the above mentioned salt efflorescence scattered by the wind, and further impregnated by saline dews, which frequently fall there during the night. Their bulk is very considerably augmented during summer, and is still increased in autumn, by the pasturage abounding in acrid herbaceous herbs. So well do these saline pastures accord with the constitution of the sheep, that in those regions they very often weigh no less than 200*lbs.*

Some of the hordes on the banks of the Volga, in the government of Casan, rear a breed of the same sheep, but very much diminished in size, both on account of the want of saline pastures, and the scarcity of winter food. Those of the Bouretes are also much diminished from the coldness of their mountainous regions, where their plants are crude, without saline impregnation ; at the same time that the country is devoid of saline efflorescence, and where even water is very scarce.

PRELIMINARY OBSERVATIONS.

In the country of Spain, where sheep are raised to a high state of perfection, part of the country abounds with very copious salt springs, and where they are deficient, their want is supplied by the care and activity of the shepherd. These springs are found in some districts, not merely in the low plains and little hills, but also issue out from some as high mountains as the whole inland country of Spain contains. In those territories where they abound, saline efflorescences are also every where to be found, and the soil partakes much of saline matter, which rises in the vegetation of grass. When the sheep derive salt from this source, it of course, in a great measure, supercedes the necessity of their receiving it from the shepherd. His giving it them, however, has been always practised both in the territories which the saline matter pervades, and in which it is altogether awanting. And of so superlative value to sheep do these long-distinguished breeders esteem it, and in so great abundance do they afford it, that "the fear of tempting the shepherds to stint them," has been assigned as "the true reason why the kings of Spain cannot raise the price of salt to the height it is in France." We shall allude to it afterwards.

Enough has surely been advanced to show the beneficial influence which salt has in confirming the constitution and in magnifying the bulk of sheep. As it can now be obtained in our country at so reduced a price, it might perhaps here, as well as in Spain fully repay the labour and expence which might be required to furnish our sheep with it; there is little doubt but it will do so. Whatever may be the manner in which it operates upon those in the countries above alluded to; —whether it acts as a preventative to diseases by which they might be reduced, and to which without salt, though unknown to us, they might be subject, or whether it immediately and directly affords nourishment, or in whatever manner it does so, I neither intend, nor do I profess myself qualified to treat. Suffice it to say, that it is either directly or indirectly productive of the most observable effects, and that the sheep which possess the advantage of it rise to a far superior value to those whose pastures nature, and the industry of man, has denied it. And such being the case, we have at least no reason to conclue that it will have no tendency to remove disease ; but on the contrary we have every reason to suppose that it will be highly instrumental in doing so. For if the sheep which with the sole advantage of the nutriment which salt affords, become so far superior to those which derive it neither from the supply of nature nor of art, we are surely justified in concluding that the former are strangers to any corroding disease, to which, for any thing we can tell, may some time or other waste the constitution of the latter, and may perhaps form the greatest barrier against their increasing in magnitude like the other. At any rate, whether or not the former are liable to disease, or to receive injury from any thing in the composition of their food, which also is a sort of disease, a check is either formed to it by the instrumentality of salt, or the hurt of which it is effective, is more than counterbalanced by the immediate fatness which the other may yield.

These observations we may now draw to a termination, by remarking the importance which should be attached to the raising of stock, both as a private and national concern. Many people not fond even of the most promising innovations, content themselves with carrying on their affairs in the old way, to which from their youth they have been accustomed, though by their backwardness they are only remaining blind to their own interest. Mixed and mutilated, and inadequate to their situation as many of the breeds of sheep still are in our country, yet the breeders of them, ignorant of the real nature of sheep, and unwilling to give in exchange the profits they draw from their present stock, for what they might possibly draw from another, continue to make no alteration, either because their prejudiced minds convince them of the decided superiority of their own, or because they conceive themselves acting upon the principle of the proverb *not to lose certainty for hope*. But every body who is at all acquainted with sheep, does not need to be informed of the superiority that a pure and proper breed possesses over one that is imperfect and improper. I would not be meant to insinuate that there is an impropriety in the mixture of any two breeds of sheep; for in this way a good and useful sheep is sometimes obtained. But to obtain this very mixture as much nicety is required, as in choosing any acknowledged distinct breed, and it is as necessary in the one case as in the other, that the holding stock be distinguished by the qualities of good sheep. So that whatever species or whatever mixture formed by the coalition of any two species, may appear to any one to correspond best with the peculiar nature of his farm, the same care and the same skill ought to be evinced in the selection of both, as also in their management afterwards.

But though for their own sakes and for the better chance of their success, it is an object of paramount importance for every person in a store-farm to give all diligence to be provided with stock as suitable and as uncontaminated as possible; it is at the same time an object of general interest. In many countries the flocks are looked upon as all the riches of the inhabitants, and in Spain they are denominated, *the jewel of the crown.* Our country is, indeed, in a higher state of cultivation than the pastoral countries of the east, and of course our lands are less appropriated to pasture; and living under happier auspices, we have the fortune to be unencumbered by the restraints imposed by the servile inhabitants of Spain, and on account of which so much emolument acrues from the flocks to the crown. Yet still in Britain the sheep form no inconsiderable part of property, and with the prosperity with which affairs relative to them are conducted, the prosperity of the state is not a little involved. In the time of the great king Edward III. who introduced into England a more salutary scheme of the woollen manufacture than had hitherto been adopted, when the wool was valued to be exported, it was then found to have brought into the kingdom £150,000 per annum, at the rate of £2 10s. per pack. And in the present improved period, when our woollen manufacture stands unrivalled by any nation in the world, and when every method is taken to prevent this valuable commodity from being transported into other countries, the annual value of wool shorn in

PRELIMINARY OBSERVATIONS.

England alone, is computed at about £5,000,000 sterling; and when manu-factured with the imported Spanish wool, amounts in value to about £20,000,000.

Every body, therefore, who has at heart the welfare of the state to which he belongs, will do whatever is in his power to promote the advancement of the concerns of stock farmers, and will give every encouragement to this branch of husbandry. And surely there are none on whom he has a more equitable claim for encouragement, and none who have it more in their power to do so, than the proprietors whose lands are pasturage for sheep, and whose interest goes hand in hand with the success and prosperity with which they meet.

A TREATISE

UPON

SHEEP, &c.

CHAPTER I.

THE DANGER OF EXTENSIVE PLOUGH-ING IN A HIGH COUNTRY.

———

PREVIOUS to entering into any discus-sion concerning the proper management of SHEEP, in the various circumstances spe-cified, it will perhaps be more advisable to treat a little of the danger and the loss attendant upon the method of keeping much land in tillage in a high district, and also of the most advantageous manner in which a STORE-FARM may be laid out and conducted.

There is perhaps, no plan that has in most cases proved more ruinous to farmers in situations, where the poverty of the soil and the backwardness of the climate, admit

A

only of Cheviot, or Black-faced sheep being kept, than that of ploughing whatever land will plough. Of late years this unprofitable plan has been by far too generally practised, and has undoubtedly contributed as a principal cause in entailing ruin upon many stock farmers. The very high price to which corn was raised during some years of the late war, was a powerful inducement to every person in the possession of land, to bring it into a state of culture. But the prospects of abundance which, in these years, gilded the hopes of the husbandman, and the avidity with which he grasped at the opportunity of participating in that abundance, operated but as the means of alluring him into that fatal security, which, to many at least, laid the foundation of ultimate misery. So great became the rage for what at that time was called improvement, and so widely, even in the most unfruitful districts of our isle, did the contagion spread, that every where—

" The messy plain, the mountain's barren brow,
Were really tortured by the tearing plough."

But the expenditure which was requisite to accomplish this important change in the sterile soil of a plain and open country, was a sum to which the farmers affixed too narrow an idea; and to meet which, the funds of many of them were far from being adequate. To bring it into any thing like a fair condition, a considerable quantity of lime was indispensably necessary; and to refund the capital consumed in furnishing this heavy article, together with payment of rent and unavoidable expences, a long continuation of the most favourable seasons, and reasonable prices, would have been little more than sufficient. But the farmers who adopted this hopeful but destructive plan, too big with splendid prospects of future wealth, did not stop to calculate upon the precariousness of the climate in which they lived, and the consequent uncertainty of crops. The great risk

which invariably accompanies the raising
of corn in a hilly region, should have been
the means of preventing all who held
farms so situated from reducing their lands
in subjection to the plough. But this
circumstance which ought to have wrought
thus upon their minds, tended rather to
invigorate them in the prosecution of their
schemes, and to excite them in making
every exertion to crop those lands with
success, for which as pasture they received but a small return, and which if
brought into a proper state of cultivation, there was at least a possibility of
their yielding produce, a great part of
which the comparative smallness of their
rents would entitle them to consider as
clear gain.   In this respect, however,
the Lammermuir farmers in general, appear to have very much mistaken the
proper line of husbandry pointed out to
them by the nature of the climate: and
by their extensive ploughing have not
only reduced their stock, to the necessity

of being kept on the most barren parts of the farm, but have profusely squandered away money on improvements which, in the present state of things, they can never hope to regain.

The experience of too many, I am afraid, will afford a convincing proof of the validity of these remarks; but that others who have fortunately been unaccustomed to such disasters, may avoid the rock, on which has been wrecked the fairest of my prospects; it may perhaps give a happier bias to their minds, to exhibit a clear statement of my agricultural concerns, for such a length of time as may enable them to form a proper estimate, of the danger of attempting to raise corn in a high district. And for this purpose, I select a period of seven years, which will afford a sufficient illustration of what I have not hesitated to assert.

I commence my statement with the year 1811. This year I had 250 acres in crop,

which, at the beginning of harvest, had
rather a flattering appearance.   But the
lateness of our climate exposes us to many
disadvantages; and is especially productive of this, in making our harvest operations generally three or four weeks later
than those in the lower parts of the
country, so that they have often their
corn brought safe into the stack-yard, before we have begun to that part of our
labour.   It is this more than any other
circumstance that gives the low country
farmers the superiority over those of a
higher district; for in point of quality in
the earlier parts of the season, the crops
frequently present 'as promising an aspect
in the one as in the other.   But the great
difficulty lies in a remoter part of the season, and it is when the time of ripening
comes that we obtain a full proof of the
danger in trusting too much to an unto-
ward climate; which scarcely in one season
out of 10 will bring the crop properly
to maturity.   The truth of this observa-

tion, I experienced in all its direful consequences with the crop of the year of which
we speak. Out of the 250 acres of corn
on my farm, only four stacks had been
secured, when the weather became so extremely wet, that all further operations
were suspended for three weeks. A large
proportion of the crop was still uncut, and
what remained in the field cut, seemed to
all appearance totally useless. When we
had almost despaired of turning it to any
account, the rain ceased and was quickly
followed by a tremendous wind, happening about the middle of October, which
completed the disaster.

That part of the crop which was cut
being in a soft loose state with the preceeding rains, was scattered over the fields
in the utmost confusion, and had to be
gathered like hay, with rakes: by which
means the straw was secured, but most of
the corn unavoidably left behind. With
respect to what was still to cut, it will
hardly be necessary to state, that, if possi-

ble, it was in a yet more wretched condition; so that of the whole crop 1 did not sell a single boll, but on the contrary had the whole seed to purchase for the ensuing spring of 1812, and also some bolls for the support of my horses.

This year I prepared for sowing 10 acres more than I had done in the one preceeding; the land was in a high state of cultivation, which gave me every reason to expect a good return. I bought seed of the very best quality from Tweed-side, which in all cost £300. But so ineffectual were my efforts, and so delusive were my hopes, that even before the seed was committed to the ground, the cold hand of misfortune seemed stretched out to nip it in the bud. On the 21st of March, there came a very heavy fall of snow, followed by a most awful drift, and was upon the whole the most stormy day ever recollected in Lammermuir, except the 25th of January, 1794.

The snow continued on the ground dur-

ing the whole of March, and a good part
of April, so that

"Winter lingering chill'd the lap of May."

The snow remained so long in the
hollows, and on the sides of ridges, that
in many places it was judged requisite to
plough and harrow it, to facilitate its
melting; and this unpromising labour
lasted till about the 7th of May, when
the seed time was finished. Nevertheless
the crop looked very well throughout
the summer months, though late. The
barley harvest was completed a few days
before the end of September. The oats,
it was thought would require ten days
or two weeks longer, until the earliest
of them would be ready for reaping.
With this view the shearers were dismissed
for the present: but misfortune was at
hand; there came such a severe frost on
the night of the 24th of September, and
the subsequent morning, that all was
ready for the sickle. For some time after

this the weather continued tolerably good, and the oat crop was safely all lodged in the stack-yard.   There was plenty of bulk, and prices being high, I still flattered myself that it might turn pretty well out.   Samples of oats and barley were accordingly prepared.   The oats, notwithstanding their premature ripening, looked very well ; they were white; and not being dry, were still plump, and had plenty of merchants.   I had my own suspicions, however, as to how they might turn out in meal, and therefore refused to deal with the country millers.   With this precaution, I sold a considerable quantity both of oats and barley, to be shipped for the London market, the former at 30s., and the latter at 39s. per boll of six Winchester bushels.   This certainly exceeded 'my most sanguine expectations, and the dark cloud that had hung over the horizon of my agricultural sphere, appeared now to be dissipated, before the bright lustre of noon-day

sunshine. The thrashing was immediately
commmenced and carried on with vigour.
The barley was a fair crop: but the
greater part of the oats in winnowing,
came out at the tail of the machine, and
had a full wind been given them, there
is not the smallest doubt, but that the
whole of them would have landed there.
My hopes were again blasted, and I
found myself in an unpleasant predica-
ment. For the article was already sold,
and the merchant no doubt looked for
the fulfilment of his bargain: and in
doing so I would have been very happy,
provided the stock should please as well
as the sample. This I easily foresaw to
be very unlikely, and was not a little
concerned. But to deliberate " in cold
debates," could promote no purpose, and
a few cart loads were accordingly sent
to the port, but alas! they were " weighed in a balance and found wanting."
From the sad deficiency which was found
in weighing the oats, the merchant caused

the whole of them to be set up until a new bargain was made.

A considerable time afterwards I attempted another sale which was happily effected with more comfort to myself and satisfaction to the merchant. The oats were sold at 29s., and the barley at 36s. With this transaction there were no reflections on either side, at least I never heard of any.

As the shepherds and other servants had to receive their dues, a few bolls of oats were made into meal for that purpose; but they produced only five stones per boll, and that very bad in quality. This was evidently a losing concern, as the oats could bring 30s. in the market, and meal could be procured of the best quality at 4s. per stone; so that by making them into meal the price was reduced to 20s. per boll. I therefore adopted the more profitable method of selling the oats and buying meal. They gradually advanced in price, and bad as mine were, some of them brought 36s. per boll.

In this way the crop paid the servants, and other necessary expenses on the farm, mostly incurred by the corn system, but left nothing for rent. For though the prices considerably exceeded the average of years, there was very little corn to the quantity of straw, as the severity of the frost had rendered a great part of it totally useless. With such a result, it is scarcely necessary to inform the reader, that the whole of the seed for crop 1813 was of necessity to be bought, and as the markets continued to rise, oats in the spring amounted to £2 per boll. Requiring 200 bolls, I had of course an outlay of £400. This year produced a very abundant crop, and prices good: being according to the fiars, I believe, 33s. 6d. per boll for barley, and 21s. for oats. Besides paying all labouring expences, and affording plenty of good wholesome food both for man and beast,

a clear profit remained more than sufficient
to pay what had been laid out the two preceding years; which gave me great encouragement to persevere in the corn system.

Crop 1814 fell far short of the preceding one. It was no doubt a pretty fair
crop, and wholesome, though in point of
quantity it was much inferior to the last
one, and prices were much reduced. Oats
and barley towards the end of the season
fell to about 16s. per boll. After paying all expences, a sum of £200 remained
in my hands, which was far below what I
had reason to expect, and by no means
adequate to meet the exigencies of such a
large concern.

Crop 1815 was a most abundant one,
but prices, as the year before, very low.
Barley at first about 17s. per boll, but
afterwards reduced to 14s. Oats also suffered a great fall from 18s. to 14s. This
crop paid all tear and wear, and left about
£300.

Crops 1816 and 17 were worse than

blanks, and like the lean kine in the vision of Pharaoh, devoured the abundance of the preceding years. In these two years I had 500 acres in crop, in a high state of cultivation, but instead of paying rent and expences, £500 of clear outlay was required to purchase corn to carry on the farming concerns.

During these seven years, then, of the respective crops of which a sufficient account has been given, there were in crop 1750 acres—which were valued at from 20s. to 40s. per acre, a small proportion of it at the highest price: and taking the average, the rent upon the whole amounted to £2000 during that period. But as the land would have been productive of corn only of very inferior value, if destitute of manure, there was for five years an annual expenditure for lime of £400, amounting in all to a sum equivalent to the rent: and which added together make a grand total of £4000.

Now, from this large sum, nothing can

with justice be deducted. Whatever money was received for my produce of hay, was all consumed in payment of servants' wages, and other necessary expences. And though the period which has been selected contains years, which were distinguished for their luxuriant harvests, yet taking one year with another, the money that I received in return for the crops of the prosperous years, was fully required to compensate the loss of the unsuccessful. So that in this attempt to acquire a fortune, all my efforts did entirely fail, and however splendid might be prospects which at one time I entertained, there remained in the end a balance against my funds of not less than £4000. Alike, I apprehend, was the fate of many, or perhaps of all, who were then similarly situated, and though the loss which they actually experienced might be far inferior to my own, yet it will probably be very near in proportion to the extent of our concerns.

It is some consolation—though a consolation that could administer but little of
the relief which in these times was needed
—to meet with a brother in affliction, who
can take up the merits of your case with
sympathy and fellow-feeling, which cannot
be properly appreciated by a person placed
in dissimilar circumstances. In these
years of general sterility, there was no
want of this melancholy kind of comfort,
but as Burns says,

> "A man may tak a neibour's part
> Yet hae nae cash to spare him."

There were not wanting however, many
instances of generosity among landlords, as
in the worst of times there are always some
to be found superior to every thing mean
and selfish. One curious case I shall relate, which will tend to confirm the remarks
that have been submitted, and will shew in
what estimation the crops of Lammermuir
were held in these years of public calamity.
In a neighbouring parish to that in which

c

I reside, a certain tenant rented a considerable farm, the lease of which was to expire in 1817. As fortune had never smiled very propitious upon this honest cultivator of the soil, so this last effort seemed to promise nothing that had been denied to his former exertions. The year preceding that in which he was to leave his farm proved more unlucky than any that he had yet numbered in the catalogue of his misfortunes. The rot and other diseases had carried off a large proportion of his flocks, and to fill up the cup of his affliction, the corn crop failed so completely, that, when he sent 20 bolls of oats to the mill, they did not produce as much meal as paid the miller for grinding them. These circumstances exhausted his funds, and he was unable to buy seed for his next crop, which caused him to apply to his landlord for aid, who readily granted him money for that purpose.

This being the tenant's way-going crop, the landlord was to be paid from the first sales made, when the crop came off the ground, but "where nothing is to be had even the king loses his right," and so also must the landlord. The crop was so bad that the tenant plainly saw it would not pay the expense of cutting, and as he was bound by his lease to leave the straw to the in-coming tenant, he told the landlord to look to the crop for his payment, and that according to the tenure of his lease he had left the straw upon the ground !! The in-coming tenant, on the other hand, insisted on having the straw ready for use, according to the custom of the country.— After much wrangling, the case was brought before the sheriff, where the landlord himself happened to be judge; but who, according to the impartiality, by which he has ever been characterised, would execute justice, though, to the injury of his own interest. In this case it is likely that some compromise had been made between the parties, as the result was never publicly known; but that neither of them would be taken advantage of, the following anecdote will shew.

The same worthy landlord had a tenant who had paid no rent for two or three terms: but at last by way of acknowledgement, his wife went to the gentleman with a brace of chickens, of which, after hearing the poor woman's story, he accepted. This circumstance the gentleman related with the greatest good humour, said he thought his tenant very honest, and believed he would have given him more if he *could* have done it.

Many cases might be produced of the folly of trusting to corn in a high country: another shall suffice for the present. A neighbour of mine in the beginning of May, 1812, passed me on my farm one morning, apparently on a journey; but he said nothing to me concerning his destination. Some days after, however, falling into conversation with an acquaintance, a farmer a little lower situated in the country, he told me my

neighbour, the person above alluded to,
had come to him in order to buy 10
bolls of seed oats, for which he was to
be paid when the other sold his hogs
in the month of June. I enquired what
might be the price, he said that as the
purchaser had a letter of credit, from a
respectable acquaintance, he had favoured
him with the oats at 43s. per boll (six
Winchester bushels.) The oats were
paid according to promise, which the
price of 27 hogs settled; but as his
land was in good order, and as the oats
were of the best quality, my neighbour
expected a good crop. A friend of mine
on his way home from Dunbar, called
upon me one day about the beginning
of September, and enquiring concerning
the state of the country in that quarter,
he said he had seen a great variety in
the state of the crop; for in East Lo-
thian the harvest was general, and the
crops good, but that my neighbour, above
alluded to, had a field of oats only in

the shot-blade.   In this backward state
in the month of September, it could
hardly be expected that the crop could
either be abundant or of good quality.
Happening to pass that way myself,
some time after the Martinmas, I was
curious to see what had become of his fine
crop of 43*s.* oats, and to my great astonishment, they were standing in the field
in small cocks, which were taken in to
the cattle as need required.

Many arguments and examples might
be produced against trusting to corn in
such a district as Lammermuir, but it
would be tedious, and, I trust, unnecessary to pursue the train any further.
Every person that considers the subject
with any degree of candour, must see
the inconsistency and the unreasonable-
ness of that system, which the above remarks have been advanced to prove.
Some enterprising minds indeed, there
may be, who notwithstanding the clearest
evidence to the contrary, may attempt to

renew, in some future period, this dangerous system. If ever any form the foolish design, it will only be those who fancy themselves possessed of superior abilities, and every way better capacitated to conduct the farming operations, than those who have before them attempted it in vain ; and their own experience, after, perhaps they have sunk a considerable fortune, will be the means by which alone they will be brought to lower the too lofty estimate which they formed of themselves, and to listen with a more willing ear to the voice of reason.

To high districts both seasons are unfavourable, the early and prolific, the late and the barren. For in the latter there is little or no crop at all; the expenses of seed and labour are all lost, and often victual to purchase at a dear rate, for the maintenance of servants whose labour has been productive of no return. And on the other hand, if at

a time, a good crop may be obtained,
then the season must have been more
than usually favourable, and a great crop
must have prevailed over the whole
country; so that even then it comes to
small account after the expenses are deducted, as the extra abundance reduces
the markets exceedingly low, and the
little profit that in these years may remain, is more than swallowed up by the
scantiness of less favourable seasons.

From all my experience I come to
this conclusion, which, indeed, I have
bought at a dear rate, never to trust to
corn for rent, in a high district, and
never to have too fine a stock. Both
are unnatural to the climate. The stock
must always be looked to as the only
certain return to the farmer, and that
should always be suited to the situation
of his farm. The only thing which
excites my astonishment is, that numbers
of the old and experienced farmers, who
have been accustomed to the production

of corn, still continue stedfastly in the
same system, notwithstanding all the repeated losses which they must often
have met. They cannot but have frequently bewailed the uncertainty of the
seasons, the fruitlessness of their endeavours, and the inconsiderable sum, which,
in an average of years, they receive in
return for their produce of grain. The
praise, of which in my opinion they are
deserving, is what in such cases can
be attributed to unshaken perseverance
through unnecessary privations and hardships. But would they only employ a
moment's reflection, and seriously bring
themselves to make a candid comparison,
between the two plans, and of altering
that to which they have been inured, I
am confident they would find cause to
repent of their past measures, and to
pursue others in which the risk and expenses are comparatively nothing, and
the return consequently much surer.

Before quitting the subject, I would

D

remark, that as the danger of ploughing, in high situated farms, beyond a limited proportion, is incalculable, and cannot be estimated by any one, but such as have had experience in that branch of husbandry, landlords would do well to restrict it to a certain quantity of acres, and not to leave it to the discretion of the tenant, who, by ill-judged ambition, may both ruin himself and hurt the landlord. Had I been restricted in my ploughing to a small portion of land, it would have been some thousands of pounds in my pocket. Experience often costs the tenant very dear, and arrived at that point where he may suppose himself entitled to some deference, he may never have it in his power to profit by his acquired knowledge. If tenants, therefore, will be so foolish, landlords or their men of business, ought to be better informed, and regulate their farming concerns by the accumulated experience of what may have happened on their

estates for the space of many years.
And it may not be improper here to
remark, that it is incumbent upon every
factor when a tenant fails or falls back
in the payment of his rent, to investi-
gate the causes of his failure, and if
it has proceeded from any injudicious
though honest ill-judged management, he
is undoubtedly entitled to more lenity
than that tenant, who, by irregular moral
conduct may have ruined both himself
and his farm.

In this way would landlords acquire
knowledge by the experience of their
tenants, and learn in future on what
conditions they ought to grant their new
leases. Instead of this however, it is
too much the practise with factors to
take the most summary way of bringing
the tenant's stock to the hammer, and
the farm to the market. And this instead of being let again to a deserving
old tenant, is often given to some new
adventurer, perhaps, almost without a

shilling that he can call his own; and, at a rent, no doubt, far above what any man of capital would venture to give. For this farce of ignorance and folly, the factor congratulates himself and his master, and exultingly exhibits the rent roll, with a few hundreds a year appended to it. The tenant, on the other hand, views his bargain in no other light than he would do a lottery ticket, and as soon as he finds the chances against him, takes every undue advantage of his farm, till in a year or two, the same scene be acted over again that had been done with his predecessor. Under such management the landlord not unfrequently loses a term's rent or two, and the very constitution of his farm being broken, he is reduced to the necessity of either taking it into his own hand for some years, or letting it at a very considerable reduction of rent to a tenant of capital, who, after the first brush of expenses is over, is in all likelihood sure to make his fortune. That such has often been the case will readily be granted by every one who has paid the least attention to what is going on, though like every other general observation liable to many exceptions; and I would not do justice to my own feelings were I to lose this opportunity of publicly acknowledging the liberality of conduct that has been practised on the estate where I have spent the most of my days.

CHAPTER II.

ON LAYING OUT AND CONDUCTING
A STORE FARM.

———

$\mathfrak{T}$HOUGH, as the preceding remarks amply testify, I would always give my decided disapprobation, to the system of cropping much cultivated land in a stock-farm; convenience, and even necessity require, that in almost all cases a few acres should be kept in culture. In doing so, however, particular care ought to be taken, that this portion be so continuous as not to be in any way separated by the intervening of pasture land, and should also be surrounded by a ring fence. It would be altogether impossible to specify the certain number of acres which should be reserved for the plough, as this must be very different on different farms, and can only be finally determined by the tenant, who is acquainted with his own individual farm. There are not many stock farms, however, that cannot, (at any rate the generality of them can,) with ease be conducted with the agricultural produce of 80 acres of land, in addition to the pasture ground. When a farm, then, is so situated as to admit, and of such extent as to require 70 or 80 acres for tillage, this would be most conveniently divided into four equal bricks: one fourth of which in rotation, to produce turnips, two fourths corn, and the remaining fourth hay. Twenty acres of fallow on a corn farm, is perhaps too large a portion for the occupation of only one plough; but as the corn, in the present case, will, or at least nearly, be consumed upon the same farm on which it is raised, carriages to a distance will seldom, if at all, be requisite, and the horses will of course have leisure to work more fallow.

But if a farm cannot conveniently afford
more than 50 acres for the purpose of
cropping, these may be most profitably
cast into three divisions, which by turns
will grow turnips, corn and hay. And if
again a still less number of acres can with
advantage be laid out in tillage, such as 40
or somewhat fewer, then it should be com-
prised in two bricks, and these to bear
turnips and corn alternately. Some lea or
meadow must then be preserved in order
to obtain hay, and if both of these are
either awanting, or inadequate to the end
proposed, the only remaining method is for
the farmer to purchase the quantity of
which he may stand in need.

Turnips and hay are by far the most valuable, as they are the most necessary crops
that can be raised in a high country. The
food requisite for the preservation of the
life of man can be procured at a season
when it is difficult and even impossible to
provide sheep with any other meat than
what is immediately at hand. In consi-

deration of this the farmer ought to look
forward to what may await him throughout
the ensuing winter, and should lay up in
store at Martinmas three months' provision
for his stock, as also for man; so that he
may be in every respect prepared for brav-
ing the storm. And in proof of the advis-
ableness of which, it will be readily recollected by many, that the severe storms
during the seasons of 1784 and 85 continued not less than 12 and 14 weeks; in
all which time the sheep could not derive
the smallest benefit from their pastures,
and could subsist only, on the turnips and
hay with which they were provided. But
many farmers little suspecting, as they
were not accustomed to, such lengthened
blasts, were not in the possession of food
sufficient for their maintenance; and the
consequence was, that in the end numbers
of their sheep perished through mere scantiness of supply.

If the few acres of arable land, which
on a stock-farm, ought always if possible to

E

be reserved, are not, and perhaps cannot, from want of stones or other inconvenient circumstances, be enclosed from the pasture, it must at any rate, lie undivided by the pasture ground, so that the sheep may not be impeded in their walk. Otherwise they will be subject to the greatest disadvantage, will frequently be tossed about by the shepherd and his dog, and in many cases, incur a damage which a richer pasture will hardly repair.

It will now be proper to turn from general observations concerning the manner in which it may be most proper to lay out the arable land of a stock farm, to the method in which the respective productions of the divisions ought to be managed. Small as the space is, which we have supposed those to occupy, the way they ought to be conducted, is not, on that account, of less importance. It is to the crops of which they are productive that the farmer has to look for the preservation of his stock in a tempestuous season, and

in a great measure also, in every season,
for the number and the condition of his
annual cast. He may, indeed, in a year
remarkably mild be enabled to winter his
sheep without the assistance of hay, and
with a very few turnips, but such years
seldom occur, and are never to be expected in a country such as this. And even in
those seasons which are peculiar for their
mildness, it is of considerable advantage
toward confirming the constitutions of the
ewes, to give them a little well win hay, in
the mornings of a black frost. Wherefore
the mind of the farmer ought in some degree to be turned to the cultivation of his
arable land, that he may allow no opportunity wilfully to escape by which he might
better the condition of his stock, and that
no storm may betake him when he is un-
prepared.

The turnips form not the least important part of his crops. It hardly lies
within our province, and indeed it is
almost unnecessary to treat of the me-

thod of rearing this crop, as almost
every body is acquainted with the manner of preparing land for turnips, and
of sowing, weeding, and bringing them
to a state of maturity.   In a high
country if the dung is good, and the
ground managed in any thing like a
proper way, it should at an average
produce 30 double cart loads per acre.
Now, the measure, which, at present, I
mean principally to recommend, is that
24 of these should be stored at, or before Martinmas.   This manner of providing for the winter, by storing of turnips is, in my opinion, worthy of every
farmer's practice.   It is of equal importance for the preservation of turnips, as
for potatoes.   Could we, indeed, ensure
ourselves of the total absence of frost,
which is a direct impossibility, or of its
delay to a certain distant period of the
season, which is equally impossible, then
surely the best method to preserve them
would be to allow them to remain on

the field as they grew, as long as there
is a continuance of mildness, and to
pull them as need may require.   But
to trust blindly to the serenity of the
weather is the most palpable stupidity,
and runing the greatest hazard; for besides inconveniences in getting them up
during the time of a storm, they are
very liable to be injured, as they are
readily touched with frost.   There are
none undoubtedly who have been accustomed to use them at table, who
have not perceived the most essential
difference in them at different times,
and felt an unpleasant sensation in eating them after having been exposed to
a piercing frost.   And it cannot be
otherwise with the inferior animals whose
palate is as capable of delicate discern-
ment as that of man, and some of them
far more acute.   Besides the disagree-
able change which with respect to taste
they invariably undergo by being exposed
to frost, turnips become liable to rot,

after this change, in which I have not
unfrequently had the unhappy experience.   In consequence of this failure
there is naturally a scarcity of meat
during the latter end of winter, and the
commencement of the spring, which fails
not to reduce the condition of the ewes
and hogs, and is the ultimate cause of
a loss of lambs more or less extensive
according to the mildness or severity of
the weather. I would not, however, be
meant to insinuate that this rottenness
is always consequent upon allowing the
turnips to be exposed to the weather,
but when left open to the inclemency of
the winter, they are exposed to a risk
which endangers their soundness, and
with them the life of the stock.   And
it would surely be a great measure of
folly in any farmer, to let this crop
lie unsheltered from the severity of the
season, when a safer method has been
devised, and without damaging the crop.
Storing of turnips possesses every ad-

vantage both in point of convenience
and safety. For besides being always
at hand, and uninjured by frost, they
may be taken out in the spring, in nearly as good condition, as those which are
standing on the field.

The Swedish and yellow kind of turnips are by nature best adapted for storing, and four-fifths of the crop, leaving
the worst, ought to be taken up, and
stored in the following manner. After
cutting off the shaw and root, taking due
care not to injure the turnip, lay them
in a heap as long as may be thought
proper; but using sufficient precaution
not to make it so wide as to exclude
the air on each side, otherwise they are
in danger of rotting. The cart being
run back to the pit and emptied, one
should be employed in laying the turnips
neatly up, making the heap gradually
narrower as it reaches nearer the top.
Then overlay them with a good covering of straw or coarse hay, roping it

tightly down, so that frost may be so far prevented from entering, as not to do any hurt. This done, you have the turnips at your command, when the snow deprives the sheep of their common manner of subsistence, alike ready in the stormy and in open weather, and beyond the reach of destruction by rot.

This plan, not altogether new, I have practised for a considerable length of time, and have found the greatest satisfaction as well as advantage in pursuing it. Before adopting it, indeed, I could seldom, if ever, bring my stock to the condition at which I constantly aimed, and they were always inferior, after all my efforts, to what, with much greater ease, I can now raise them. As far as my observation and experience extend, it is the most profitable and advantageous method which any farmer, in a high district can follow; and I doubt not will be successfully practised by all after the experience of a fair trial.

If the measures above recommended meet with the approbation of others, after their turnips are stored, there will remain in the field about a fifth of them, and these the most inferior in quality. They are reserved as a portion for the hogs during the earlier part of the season; but the manner in which they ought to be eaten, shall be hereafter more particularly adverted to, when treating of the Rearing of Sheep.

We have now terminated our few remarks respecting what concerns the produce of one of the divisions of the arable land upon a stock farm. The crop which stands next in importance, or which is perhaps of equal value to the stock farmer is that of hay. The manner of managing this crop is so much alike in every district, and no new innovations having been attempted in this branch of husbandry, it will be almost unnecessary to dwell upon it. After being cut, won, and put up in the

usual way, the farmer can only cause it
to be brought to such places of his
farm as may be considered most convenient, and reserved for any storm of
which the season may be productive.
In the event of which, the manner in
which it ought to be distributed for the
support of the sheep, shall be taken
notice of in a subsequent page.

Should any hay, however, remain un-
consumed from a former year, it will
acquire an unpleasant taste, and will not
be eaten with desire by the sheep. That
this may be accompanied with no loss,
when it occurs to any extent, the new
hay ought to be stacked as green as
possible, and mixed with the old, by
which means both will be rendered agree-
able.

When the arable land in any farm is
of so limited a size, that it cannot possibly
afford any for the production of clover
hay, or if its disadvantageous situation
render it extremely inconvenient, other

measures must of necessity be had recourse to in order to obtain it, as no
stock farm can be at all conducted, with
the least probability of success, without
that commodity. Perhaps in a case of
this sort, some corner, by allowing it to
lie during the former part of the year
unoccupied by any stock, may possibly
raise a quantity of natural hay, which will
supply the deficiency of the other kind.
If this hay be properly won, and have a
fine flavour, the sheep will eat it with
more fondness than even clover hay;
and if it be the produce of a loam soil,
their fondness for it will be considerably
increased. When this also cannot be procured, meadow, or as it is generally
termed bog hay, is sometimes found
pretty good; but always inferior to the
natural lea hay. If the field allotted for
the purpose of obtaining either of these
kinds of hay be not quite enough prolific,
overflowing it with water will add not a
little to its fruitfulness; and if that can-

not be done, spreading it with a sufficient quantity of dung will also have a good
effect, which can be easily procured by
reducing a little surface ground to ashes,
if not by other means.

One other sort of produce to be raised
upon the arable land of a stock farm
only remains to be considered. If plenty
of land can be obtained, the corn should
occupy as much as the turnips and hay
together, that is two-fourths, but may be
varied with respect to extent as the farm
is of larger or smaller size. Though
this crop to a stock farmer, is of inferior
value to any of the other two, yet a
certain quantity of it is in the highest
degree requisite, and cannot well be altogether wanted. Were it not indeed for
the sake of turnips and hay, it would
perhaps be more advisable for the farmer
to let the whole of his land lie in the state
of nature, and purchase what corn he
needed. But as the two former crops
can by no means be dispensed with, that

the divisions of the cultivated land may be regularly cropped, corn must also be raised in rotation. Though comparatively fewer hands are required to the management of a stock, than of a corn farm of the same extent, yet these few stand in need of the means of comfortable subsistence as well as if they were much more numerous. A little corn will therefore, be absolutely necessary for the maintenance both of the persons employed in carrying on the farming operations, and also for the master's own house and stable. Each shepherd and his family will consume oatmeal at the rate of about a stone per week, and this, together with what may be used in support of the master's own family, domestics, horses, &c. will not require so little as may be supposed, were it all to buy. It will always be found to be attended with less expense and with far more convenience, to grow at least as much corn as will be destroyed upon the farm. And

should any remain after all demands are
satisfied, it will in most cases be useful
to have some bolls to expose to sale (as
the markets for stock are so few and so
distant,) that the wages of servants and
any small expenses necessarily incurred
may be readily defrayed.

Should any difficulty at times be found
in providing dung for fallow, the same
method must be had recourse to, which
was mentioned in the failure of hay,
that of burning surface ground.   To
supply the deficiency of manure in this
manner, the easiest way in which it can
be done is, by ploughing the surface of
a piece of coarse ground with Mr Fin-
layson's Rid plough, and afterwards col-
lecting it into heaps, and burning the
earth thus procured.   Plenty of time
will always occur for this business during the period that intervenes between
the finishing of the turnips and the commencement of harvest.   And by the a-
bundant application of these ashes to the

land, a good crop of turnips in ordinary seasons will be insured.

This brings to a close our remarks upon the manner of laying out, and conducting the arable land belonging to a stock farm. The advantages which the narrow scheme above recommended, may possess above the more extensive one of bringing into cultivation a great part of the farm, will be greater or smaller according to the state and situation of the different farms on which it may be adopted. There may be little difficulty in finding some containing a great proportion of stock land on which corn has been raised with tolerable success. But such incidents are rare, and ten perhaps may be placed in opposition to one, in which it has altogether failed; at least been so unprofitable that stock might have been kept with much greater advantage. It will be observed that what I say refers only to farms in Lammermuir, or to districts equally barren and hazardous, and without entering

into any further detail concerning the truth
of the above remark, I would merely ap‑
peal to the experience of those farmers
who have changed the method of conduct‑
ing their farms from attempts at growing
corn, to the manner already described,
whether they have any wish to return to
their former system ? While to those who
persevere in their ploughing, I would sim‑
ply ask them to calculate the profit which
they could gain by stocking that farm which
they have in tillage, and then to recur to
what, in an average of years they have
received for their productions of corn,
after deducting all expenses in manuring
the land, loss of seasons, and other una‑
voidable circumstances ; and after a fair
comparison of necessary expenditure and
annual gain on both methods, I am quite
certain, in most cases at least, that the ba‑
lance of profit will arise, not from corn,
but from the keeping of stock.

But the most extensive part of a
stock farm still remains to be treated

of. Though more extensive, however, it
will hardly admit of as much discussion.
In such districts as that of Lammermuir,
the pasture land is generally of inferior
quality, and but a small proportion of it
is made up of grass. But poor though it
be, it differs no less in degree than does
that of corn. This circumstance renders
it necessary for the stock farmer to consider with what kinds of sheep his farm
ought to be stocked. I allude not at
present so much to the different breeds
of sheep, as to the different ages and
kinds of the same breed. By inattention or ignorance in this respect, it is
possible enough that loss to some extent
may soon be felt. Should, for instance,
ewes be put upon ground which is capable
only of maintaining hogs, this mistake
will subject the farmer to a loss of profit
which he might have averted by a more
skilful management. It is customary, I
am aware, in the Etterick forest to allow
the hogs to pasture with their mothers,

G

and to graze upon the pastures where
they were nourished antecedent to weaning time.   This it is thought, is of no
small benefit toward preventing the sickness.   But whatever may be the advantages or disadvantages which it possesses,
and however well it may be adapted to
some farms, there are many on which
it could be practised not only with no
profit, but would also be followed by
an unfavourable result.   And we may
almost lay it down as a general rule,
that whenever any farm, or part of any
farm, contains soft rough ground, this
would with most advantage be appropriated for hogs: and, on the other
hand natural bare land is more suitable pasture for ewes than for hogs.

But be the stock what it may, in
farms of large dimensions, it must be
divided into different lots, or, as they are
commonly called, hirsels, varying in number and size according to the particular
nature of each farm.   The pasture allot-

ted for these respective hirsels, should as much as possible partake of the same quality. For if there be in it a mixture of fine and coarse, the sheep will derive little or no benefit from the latter, and will always have an inclination to run to the former. This, it is evident, cannot fail to be productive of injury to them, and to lessen in no small degree the good which they would derive from the more valuable ground. To avert such an evil, the land ought to be so proportioned, that the inferior pasture exceed not in extent a fifth of the grassy soil. This much, as affording a change, might yield some advantage to a hirsel, but more than this would, at least be useless.

But should a large tract of inferior land lie adjacent to what is of superior quality, the one should be altogether separated from the other by a temporary fence, and pastured by distinct hirsels, if necessary by different kinds of sheep. By this measure it will not only support a greater number than it would otherwise have done, but will also keep them in much better condition. This observation has been confirmed by repeated experiments made in the course of my grazing concerns.

A subject deserving of much more weighty consideration, however, from the stock farmer, is the attention which he ought to bestow on having his farm well provided with shelter. In high districts this is frequently in a great measure supplied by the works of nature. Ranges of mountains, and hollows surrounded by ridges of hills, are to be found in most Highland farms, which are of inestimable benefit for the preservation of the sheep in winter. There is no doubt a danger attending a storm, when they are lying in the covert of a hill side screened from the severity of the blast, of their being overwhelmed under an accumulating heap of snow. If, indeed, in a heavy fall ac-

companied with drift, they be lying near
the brow of the sheltered side of the mountain, they are sure of being suffocated
under the pressure of an insupportable load;
but if towards the bottom, provided the
hill be sufficiently steep, they will rest in
a place of safety whilst the storm scowls
above their heads. Much loss has, however, been sustained from the circumstances of their being improperly situated
on the hills, and also, in many cases, from
these being inadequate completely to de-
fend the flocks in time of danger. In
consequence of this, the 25th of January,
1794, proved fatal to thousands, as have
several other years, though seldom ever to
such a dreadful extent. The misfortune
is that many trust too much in this sort of
shelter, and expose their stock to a risk
which a very moderate expense would
divert. For however well some farms may
be provided with natural, there is almost
always a need, more or less, of artificial
shelter; even in those farms which have

every advantage from the fortunate position
of hills, a limited provision after this manner is safer, and will generally be of much
use in a storm.

Undoubtedly the best artificial shelter
that can possibly be made, is by strips or
clumps of plantations.   But on account
both of the expense attendant upon finishing these in a proper manner, and the
time which the trees necessarily require to
come to any degree of perfection, it can
never be advisable for a farmer to store his
farm with these out of his own funds.
Shelter of this sort should in every instance
be afforded at the expence of the proprietor, and is an object worthy the most
serious attention of every landlord, as in
all high farms it would increase the value
of his property nearly at the rate of 10 per
cent. The manner in which they ought to
be constructed, that they may be found of
equal value from whatever quarter the
storm rages, is to plant with spruce and
larch firs an extent of from two to four

acres at suitabledistances, and with four
sides each in the form of a crescent or
half circle bending inwards. This, if
properly executed, is the most valuable
shelter that can be procured upon a stock-
farm after the trees have arrived at a
sufficient length.

But important as this acquisition would
be both to the proprietor and to the
occupant, there are many farms that are
utterly destitute, or rather there are few
that can properly be said to possess them;
and since this is so much neglected by
the landlord, it is the object of the tenant
to employ such means as he has in his
power, for accomplishing the same end.
The easiest method to which, in these
circumstances, he can have recourse, is
the erection of stells. These can be
built at a very small expense, and can be
wanted upon exceeding few stock farms
only, if upon any, that are unfurnished
with plantations. But in cases where
little or no advantage arises from the

situation of hills, a considerable number of them must be built to be productive of the good intended, so that sheep may not be to drive to any distance at the commencement of a storm, which in many instances could not be done. A calculation has been made in a late work by an ingenious author ;* though, with all due deference to the honourable gentleman, I must state it as my opinion, that the number which he considers as necessary, are more than sufficient. There are

---

* A Treatise upon Practical Store Farming by the Hon. Captain Napier. The great advantage of stells is shown by powerful facts in a chapter upon "stells and storm feeding." In a letter inserted in the above work, subscribed by a shepherd, Alexander Laidlaw, the convenience and the large profit accruing from stells, is clearly stated by a comparative examination of the loss and condition of the stock upon two neighbouring farms, the one well provided with stells and hay, and the other in a great measure destitute of both. The decided superiority of the former is proved in a clear though rather sarcastical manner by this Etterick Shepherd.

many farms, even the generality of them, will have food quite enough to support, taking into account both good and bad land, 1000 sheep upon 1200 acres. Now it does not appear to me, that on the approach of a storm, there could be much inconvenience in getting 1000 sheep into 16 stells, which is one to every 75 acres of land. And that if they are properly situated, and the shepherd diligent in the performance of his duty, the blasts of winter will arrive without any deadly consequence. It is, indeed, better to err on the safe side, but in general there can be little risk at that proportion, and if the expense that would be required to erect an additional third, can with propriety be avoided, the farmer in these times, at least, would do well to retain it. But as I would differ only with reluctance from the opinion of the author above alluded to, I shall leave it to the private determination of every farmer what number of stells he may consider most

H

proper, as it may be somewhat varied, according to the peculiar nature of each farm.

In place of stells sheepcotes have sometimes been substituted, to which I must give my decided disapprobation. These are far too confined and warm for sheep, and make them unwilling to go out in days when they might derive much advantage from their pasture. Stells possess the superiority in every respect, and ought to be constructed, neither in a square nor circular form as is most customary, but in the form which I shall describe. Every stell ought to occupy at least half an acre of ground. Similar to the plantations it should be constructed with four sides bending inwards in the form of a crescent or half circle. A dyke should also be brought from every corner, and continued for the length of 10 yards or so. If this is to be done at the farmer's expence the first three feet may be built with stone, and

other three with good substantial sod;
which, if done by contract, should not
cost more than 2s. per rood.

The sheep, however, ought not in a
storm to lie in the area of the stell, as
is done with those which are of common
construction; when in these cases, indeed they are in the inside, there is
little danger of their being blown up;
but whilst the snow continues falling,
and when it is drifting, there is such a
suction in the stell that they cannot remain there with any comfort. This circumstance has not surely escaped the
observation of any who have had occasion to witness sheep in stells during a
stormy day. The advantage of the particular structure which I have recommended, is so far good, that it is free
from this disadvantage. Instead of putting the sheep, therefore, within the stell,
they ought to be laid into the circle
of the opposite side from that of which
the wind blows. The two dykes and

area of the stell, detain the snow so
completely that the outside, where the
sheep should lie, must remain quite clear,
and they will consequently rest there un-
exposed to the severity of the weather.
And if this plan be adopted, the provision of hay for convenience' sake, may
be stacked in the beginning of the season within the stells, and not at the end
as Captain Napier recommends.

It was at one time a very common,
but, I believe, now an almost abandoned plan, to keep up a stell or two for
the nearly exclusive purpose of holding
the sheep throughout the course of the
night. The advantages arising from this
manner of folding the sheep, as it was
termed, are too trifling ever to be compared with the great injury which they
sustained by it. That it is now left off
being practised is a proof that the nature
of sheep is better understood, and to a
person possessed of the most limited
knowledge of them, it would be super-

fluous to point out the defects of that
method.

It may also be remarked in treating
of shelter, that patches of whins or furze
ought to be sown where the soil is
adapted to their growth, and of broom
too ; the former affording food as well
as shelter in a storm. Besides these
there are other shrubs and plants which
ought in some places to be cultivated
as antidotes against disease as well as
for food and shelter.

In an open stock farm it will always
be found requisite to have a park or two
in reserve for diseased sheep, tups, &c.
Four are mentioned, by the honourable
author lately referred to, as necessary, and
for the various purposes of a " lambing
park, a hay park, a twin and tup park,
and an hospital park." If the system be
adopted with respect to the raising of
hay in rotation with other crops, which
was pointed out towards the beginning
of our observations, the necessity of a

park for that purpose will be superseded. And as the invalids may be classed either with the twins and tups or with the young lambs, it seems to me that two enclosures will be sufficient. Perhaps the tups may graze during the greater part of the season with the hirsels upon the best pasture: and as any common park dyke will be too low to confine them, towards the latter end of the year, they might be put into stells for a month before they are to be used, and fed with hay and turnips.

With another remark we quit the subject, and conclude by recommending the method of surface draining, as a very effectual one for improving wet ground. For this kind of land, it is the most important improvement that can possibly be made, and wherever it has been performed, its happy consequences have always been felt. But as this is one of the most effectual means for preventing the *rot*, we shall defer treating of it at present.

CHAPTER III.

ON BREEDING CHEVIOT AND BLACK-FACED SHEEP.

———

THERE is perhaps no department to which the attention of the stock farmer ought to be more carefully directed, than to the breeding of sheep. For however skilfully he may conduct his other farming concerns, if he considers this as subordinate and inferior to the rest, he will fail in obtaining a desirable stock. Yet notwithstanding its manifest importance, the regard which has generally been bestowed upon it, is far from being commensurate to what it deserves. There is indeed an observable improvement in this, as in almost every other branch of farming; but there is still much room for further progress, and to heedlessness in this respect may still be traced, in a great measure at least, that deficiency in point of beauty and usefulness which strikes too forcibly upon our notice in the general exhibitions of stock at our public markets. Since, therefore, it is a subject which is apparently not sufficiently understood, and as the Border farmers by whom it seems to be more fully comprehended, and more practically attended to, have not furnished the public with their opinions concerning it, it may not perhaps appear improper that I have laid hold on the present occasion, of submitting the few cursory remarks which my observation has collected on breeding Cheviot and Black-faced Sheep.

The first great object that demands the attention of the farmer who wishes to be successful in breeding sheep, is to make the specific breed which he possesses, correspond with the respective nature of his farm.   For as certain constitutions only are fitted to inhabit climates of a certain temperature, so also there are different kinds of sheep which are by nature constituted to subsist in different districts.   A large catalogue of distinct breeds have been enumerated, each as possessing some quality not common to any other, but the three to which the pastures of this part of our island are almost entirely confined, are the Dishley or Leicester, the Cheviot, and the Highland or Black-faced.   These differ very widely in their constitutions ; the first answering only the low-land, the second the mid-land, and the third the high-land districts.   And it is at once evident to every one who is in the least acquainted with what is peculiar to these breeds, that it would be a glaring absurdity to transfer the Leicester breed, naturally fitted for mild weather and fertile fields, to a higher region, where they would experience an unaccustomed severity of

I

cold, and where the soil bears the stamp
of barrenness and poverty. It is also evident that the sheep inured to a cold
climate and unprolific pastures, would
produce comparatively small profits in
luxuriant fields, to what would arise from
the large growthy sheep, which with the
same food could be brought to a far
superior value, and which acquire for
their support our finest pastures. A
certain degree of discrimination, it is,
therefore, necessary to observe, in choosing what breed of sheep is best suited to
the peculiar state in which the farm, with
regard to soil and climate, may be situated. For if any one, without due de-
liberation, proceeds to stock his farm
with such sheep as are not fitted for his
soil and climate, he will in all probability
very soon feel the heavy consequence of
his inexperience ; especially if he attempt to keep those for the support of
which his pastures are incompetent. And
in this case, with whatever diligence and

activity he may provide every convenience and every advantageous circumstance that may tend to better their con-
-dition, he is only making an effort that
exceeds his strength, and the likelihood
of his success would be much greater
were he to confine his endeavours to a
level with his capacities.

Of the truth of this, my own affairs
have unfortunately furnished me with sufficient corroboration. The farm which I
at present occupy, has been rented by
our family for nearly half a century.
Upon entering it at first, the Cheviot
stock was the object of our choice;
which species was selected on account
of the farm being situated in what may
be called a mid-land district. So long
as we continued in possession of this
breed, every thing proceeded in an even
manner, and with considerable success.
But in a time when almost every body
was in admiration, and if possible in possession of the Dishley or Leicester breed,

we also, influenced by the same spirit, conceived a distaste for the Cheviot, cleared our farm of them, and with more flattering prospects, as we supposed, procured the more fashionable stock. Time, however, convinced us of the mistake in the most decisive manner; our coarse and lean pastures were unequal to the task of supporting such heavy-bodied sheep, they gradually dwindled away into less and less bulk, each generation, was if possible, inferior to the preceeding one, and when the spring was severe, seldom more than two thirds of the lambs could survive the ravages of the storm. The ewes, indeed, fed well, but could never exceed the small weight of 12*lbs.* or 13*lbs.* per quarter. A manner of conducting the farm, so unsatisfactory and so unprofitable, was, after some years, abandoned as fruitless, and I formed the resolution of stocking it anew with the Cheviot breed, which I got from a distinguished breeder on the Border. These,

as formerly, correspond entirely with the
nature of the farm, feed with the greatest facility, to a weight surpassing that
of the former stock, not less than *3lbs.*
or *4lbs.* per quarter, and bring up lambs
at least equal in number to the ewes,
except in the most disastrous seasons.
Some, however, still remain blind to this
manifest advantage, and here, as in the
corn system, I am not a little astonished
to see a few farmers, even in my own
neighbourhood, retaining to this day the
Leicester breed, though it has degenerat-
ed in the most obvious manner, to a
paltry, trifling size, and though one
hour's reflection, upon the comparative
success of others, might convince them
of their error. But these, I am afraid,
are too wise to receive instruction, and
I would advise such as are willing to
take advice, and would consult their own
advantage, to learn from the affairs of
others, not hastily nor with premature de-

termination, to stock their farm, without attentively regarding its situation.

But this, though not the least important, is not the only circumstance worthy of the special care of the breeder. The qualities of the sheep which he selects from a particular breed ought to be no less the object of his attention. For if the original stock, from which, he is to raise his annual cast, be inferior to what he wishes their produce to be, he will never attain the end he has in view. His first care ought to be to examine with minuteness into their form, and provide himself only with those, in which he can trace the lineaments of good and well proportioned sheep. And with the view of affording some aid toward the accomplishment of this, it may not be unnecessary to give a general description of the properties of a good Cheviot sheep.

This breed is "hornless, their face and legs in general white; the best kinds have a fine open countenance, with

lively prominent eyes, and body long;
fine clean and small boned legs, and thin
pelts." They ought also to have a
large ear, and to be long from the ear
to the nose. The true kind are well
proportioned in their quarters, and have
a good thick cover of wool extending
over their whole body. It ought to
come well forward behind the ear, but
not at all to reach over the face. The
mutton and wool should likewise fall
well down toward the knee, and although
the wool is, and should be, rather coarse
upon the thigh, that is productive of no
loss to the farmer. The deficiency in
point of quality, is fully compensated by
the abundant growth which takes place
upon that quarter. This circumstance
also renders the sheep better fitted to
withstand the cold weather and rough
blasts peculiar to high districts.

One other distinguishing mark of good
sheep respects their countenance when
lambs. Their eyes and ears should then

be discriminatingly examined, and such
as are red may be considered as strong
indications of a weakly constitution. The
lambs that are stamped with these unfavourable marks are always the most
delicate, and if able to escape the hazard
of the spring, come to smallest account
in an open country.

The selection of tups is of the highest importance, and of late years appears
in general to be more particularly attended to. They should of course be
possessed of every mark which is expressive of beauty in a sheep. To them
as well as to ewes may be applied the
above short description. Farmers ought
to be especially careful in examining
whether they have a close coat of wool,
as a deficiency in this respect will hardly be overbalanced by an assemblage of
other good qualities. They should also
be full behind the shoulder, have a
long straight back, round in the rib, a
clean face, and full of action.

The exact period at which tups ought to be put to the ewes cannot be altogether determined, as this may vary according to the situation and circumstances of each farm.   The variation, however, in almost every farm, where the Cheviot breed is properly kept, is very small, and the time at which the tups ought to be let in amongst the ewes may perhaps be restricted to the days between the 15th and 22d of November.   Whether about the first or the last of these days may be chosen, as most suitable for the respective nature of different farms, a few days longer should always be allowed to elapse, before they are put amongst the gimmers.   The advantage arising hence is, that the latter being less able than ewes to endure the hardships of lambing and of giving suck, should have a little longer before the commencement of their lambing season, that the weather may become somewhat milder, and the pastures beginning to yield more nourishing food.

The proportion of tups to ewes will, in almost every case, require to be very nearly the same. One tup will generally be found quite sufficient for three score of ewes, and to lessen the number of tups below this proportion, will always be found dangerous. But if the hirsel contain a larger than common number of ewes, and the pasture on which they graze more than ordinarily steep, this proportion will probably be too small, and can only be properly determined by the experience of the farmer. From this number, however, I never suffered any loss.

There is a measure concerning the tups that I would here recommend, which is, not to retain the same ones for any length of time upon the same farm. That the contrary practice is productive of any hurtful consequence is, indeed, disallowed by some, but from

my own observation, I am rather inclined to think, that when they are
continued from season to season without
alteration, the breed gradually degener-
ates, and becomes more and more weak.
To prevent the stock from incurring any
injury at all on this account, one half of
them should if possible be annually exchanged, provided other ones of superior or of equal value can be substituted in their room. If this cannot be
done; it would certainly be folly to
part with better ones for worse, merely
for the sake of making a change. But
some, however few, should, if possible,
be yearly exchanged.

With respect to the breeding of the
black-faced sheep, what has been advanced concerning Cheviot, is, with little alteration, also applicable to them. Their
form, however varies much from the
Cheviot. They have for the most part
horns, black faces and legs; "a fierce,
wild-looking eye, and short, firm, hand-

some carcases, covered with long, open, coarse, shagged wool." This breed is undoubtedly better adapted and more profitable than the other species for mountainous districts. None have ever yet appeared of a constitution so hardy, and so favourable for the highest grounds in our country, particularly where that is covered with heath. With this breed there is generally little loss in lambing time, when compared with what usually takes place among the Cheviot; and they are much easier maintained when hogs. Their wool being exceedingly coarse sells always about a third below the price of Cheviot. But the weight of the former is somewhat heavier than the latter, bearing to each other the proportion of about five to six. The tups of the black-faced breed are commonly let to the ewes about, or a little after, the 20th of November: one tup as with the Cheviot serving for three score of ewes.

In breeding from black-faced sheep,
they ought never to receive a tup of
a different breed, either of the Leicester
or Cheviot. A good sheep is never
produced from their being crossed, but
is always ugly and ill-shaped. From
Cheviot ewes, however, and a Leicester
tup, a very good, well-made sheep may
be obtained; it can be raised to a great
weight, and well fitted for the butcher.
A mongrel breed thus formed is indeed
very seldom bred from again; but in
my opinion it may, in some places, be
attended with as much, even with more,
success, than any other breed. There
are many situations that are rather too
highly situated for keeping sheep of the
Leicester breed, and are perhaps more
than qualified for those of the Cheviot,
on which they may be reared to great
advantage. And as there may very probably be some difficulty in finding a
good market for the ewe lambs, they
will be most favourably disposed of in

the butcher market as fat. If the wether lambs can be continued till they have
become dinmonts, they will be accompanied with more profit, than if they are
sold when lambs.

CHAPTER IV.

ON REARING CHEVIOT AND BLACK-
FACED SHEEP.

———

IN treating of the rearing of sheep,
we shall commence at the period of
their separation from their mothers, and
trace them through the various stages at
which they successively arrive, till they
have reached a state of maturity. In
pursuing this natural course, I may, perhaps, have occasion to recommend measures which it may not be expedient
for some to practise; but these may be
considered as exceptions to the general
rule, and, avoiding every thing of a particular nature, I have endeavoured as
much as possible to state only what is

applicable to the above-mentioned stock in high situations, which is their proper sphere. Some of the remarks contained, however, have only a reference to sheep belonging to the Cheviot breed alone.

As there are few or no natural diseases incident to lambs, the principal object of the farmer is to employ every means in his power to free them from external danger, and to accomodate them with a sufficient supply of milk. As the gimmers, with the same treatment, will generally be unable to nurse their lambs to so much advantage as the ewes, that they may be brought to something near an equal footing to them, and that they may be the better enabled to bring up lambs capable of being classed with those of the latter, they ought to be separated from them about a month before the commencement of their lambing time, in order to receive turnips. These should be daily laid down to them at the rate of

a double-cart load to every five or six
score. This additional supply of food,
will assist them in undergoing the hardships which as mothers they have to endure ; and this much will be found absolutely necessary to qualify them for
supporting lambs of equal value to those
which are nourished by the ewes.

At the end of this period, and when
they are beginning to lamb, they should
then be re-joined to the hirsel from
which they were taken, and eat turnips
in common with the rest. The stores
of these may, indeed, by this time be
almost exhausted, but there should still
remain a sufficient quantity for the
whole hirsel to receive them, during
some length of time, at the allowance of
a cart load to each eight score. If
theirs be a hilly pasture, the great advantage of turnips should be prolonged
to the same extent throughout the season of lambing time, and should be
given them in a place either naturally

L

or artificially sheltered.   Perhaps in a
mild and early spring the continuance
of them might be somewhat abated ;
but such happening very unfrequently in
a high country is never to be depended
upon ; and as it is the duty of a farmer
to make provision, not for a mild, but
for a severe spring, when he knows the
latter may very probably prevail, the
turnips for which he has wisely provided, and part of which he might want
in a season of the former kind, will not
be mispent when given to the ewes a
little longer than absolute necessity
might require.   Upon the whole, this
method as above recommended, will be
found in high districts the most profitable in the end, as it will be the means
of preserving alive a great number of
lambs which, in unfavourable years,
would, but for turnips, have been swept
away in the blast, and of raising them
to a condition, which, but for them,
they could not have attained.

The lambing season is above every
other to the stock farmer, the most important.   It may very appropriately be
denominated his harvest, and in it as in
that busy period, he should approve
himself more diligent than in any other.
To the shepherd also it is a time of the
hardest trial, and during which he is
entrusted with the heaviest charge.   But
his duty exceeds what he has ability to
perform, if he be entrusted with the
care of more than 400 sheep, and in
some cases, perhaps, even with a more
limited number than this.   And when
this bound is overstepped, as it occasionally will, one shepherd being more
than able to tend in many places, during the rest of the year, a larger hirsel, he ought to be provided with an assistant, in this perilous time.   Nothing
more should be required of him than to
fulfil unassisted his office during the day,
whilst another should be entrusted with
the execution of it in his absence du-

ring the night. It is then for this additional person to watch them with the care and attention which the occasion demands, they being put, if possible, into a convenient place of shelter, by the shepherd himself, on the close of each succeeding day. Hither should the assistant oftentimes repair throughout the night, visiting it at the distance perhaps of every two hours, to assist any ewes that may require help in bringing forth, or to carry any weakly lambs into a house ; one for which purpose should be prepared at hand. The lives which this man might spare in one night alone, might more than discharge the expenses which his attendance would incur.

Whatever precautions may be taken, however, some cause unforeseen or inevitable, will, unless in extreme cases, in spite of every effort, deprive certain ewes of their lambs ; in which case it will be proper to substitute another in its stead. If the ewes are in any thing like, what may be called, good condition, there will in all probability be as many twin lambs, as will supply the place of those that suffer by death. In some instances the mothers that have been deprived of their own lambs, will take another with astonishing fondness; but should any difficulty be found in this, skinning the dead one, and covering with it the one that is to be substituted in its stead, would be attended with a good effect; shutting both up, at the same time, in a small dark corner, for the space of 24 hours. This method will convert the most stubborn aversion into attachment; but the latter measure, of confining them in any small corner for the length of time specified, will generally be found successful without having recourse to the skin of the deceased lamb.

I have already had occasion to touch upon the measure of a park being reserved for the purpose of containing the twin lambs. There are, no doubt, many farms on which this would be kept up with advantage, and on which it could not with propriety be dispensed with. But in a high district similar to that of Lammermuir, any enclosure for that purpose will be found totally superfluous. So far from any thing of this kind being needed, it is accounted very fair, and beyond which even the hopes of farmers in situations such as this seldom extend, if they can, by every endeavour on their part, bring up to be weaned lambs equal in number to the ewes. Twins, indeed, there may be, but these are generally fully, and often more than fully, required to compensate the loss unavoidably sutained by the ravages of storm and disease. And I am certain the experience of highly situated breeders sufficiently proves, that the necessity of a twin park, is too truly superseded, both by the real scarcity of twins, and the want produced by occasional deaths.

After the season of lambing has elapsed, the first circumstance of importance that regards the treatment of the lambs, is that of gelding the males. This operation, though exceedingly simple, should be proceeded to with great caution. The operator should, by all means, abstain from spiritous liquors of any sort, and the lambs lifted with as much gentleness as possible. The knife with which the operation is performed ought to be sharp and smoothedged. To prevent any death by mortification, it was once a common custom, and by some farmers is still retained, to anoint the wound with turpentine. This, as Mr Hogg expresses it, is a sure, but a severe remedy ; having such an injurious tendency, that no less than 14 days are requisite for the recovery of the lamb. This is, indeed, a terrible preventative, and every means should be tried to render unnecessary so hurtful a remedy. The danger, I believe, is in some measure dependant upon the condition of the lambs themselves, and the peculiar nature of the ground on which they pasture.

When they are fat there is more to be dreaded : what pasturage is unfavourable, experience will best determine. If the accompanying circumstances, however, are duly attended to, and the operation itself performed with sufficient caution, there will be little damage sustained upon any farm; less than is generally experienced by the application of turpentine. The day, in the first place, should not be finally resolved upon long before the measure is intended to be put in execution, as the weather in a few days may undergo a considerable change. If the atmosphere be sultry, it is an unfavourable season to geld lambs : and when such is the state of the weather, that business ought to be postponed until it is again purged of electrical matter. But more danger is consequent upon the lambs being heated to any excess. If great care in this respect be not taken the most deadly effects will not fail to ensue, as to geld them when they are violently heated is

the sure engine of death.   They should
be put into a fold, erected for this and
other similar purposes, the night preceding, where they may be ready for the
operation being performed at an early
hour the following morning.   This fold,
or whatever place is assigned for that purpose, should also be carefully prepared,
cleaning it from every kind of foul dirt or
nuisance that may in the least tend to in-
flame the incision.   If these precautions
are all attended to, and if the operator
be acquainted with the proper manner of
procedure, few deaths, if any, will succeed.   During the practice of many
years, I can hardly say that ever I
have suffered any loss on account of
this operation, neither will it be otherwise upon most other farms, if properly
conducted.

Immediately after this operation another
one of inferior importance and beyond
the reach of danger, may be carried on,
that of ear-marking the lambs.   A person

M

should be appointed for this exclusive
purpose, that both may be done before
they are again placed upon their feet.
Two instruments are necessary to accomplish it aright, the one making a circular,
the other a triangular hole. One of these
may be appropriated for marking the ewe,
the other the wether lambs. The marks
may be varied from one ear to another,
and to different parts of the same ear,
that the distinction may be fully kept up.
This measure is only thus far of consequence, that it serves to distinguish the
ewe from the wether hogs, should any
accident mingle them together, or in cases
where they graze in one hirsel. And as
it is necessary to know the respective
ages of the ewes to determine which are
to be sold for draft, the ear-mark will be
the simplest and the most decisive method of becoming acquainted with, provided care be taken to remember how
the lambs are marked every year. Or
should they be carried away by stealth,

or accidentally stray to an adjoining farm, the ear-mark will be further of use, as it will furnish an additional proof of their identity.

Betwixt this and weaning time, nothing occurs of which it is of importance to speak. During this while nothing lies in the way to interrupt them in their regular course. If they are to be continued as hogs, the period at which they are commonly weaned is near the middle of July; as, they ought always to suckle three months. This length of time under the nourishment of milk is essentially necessary to confirm their constitutions, and to lay a steady foundation for their future increase. It is not likely, however, that a farmer in a high district will be able to wean a lamb for every ewe. There are generally some ewes that have no lambs at all; and what from this, the inclemency of weather, and adventitious circumstances are awanting, will generally be found to lessen the proportion of lambs

to 19 for every score of ewes. It is the
object of improvements to preserve at
least this number, but where these are
overlooked a much greater short-coming
may be expected.

With respect to the treatment of ewes
upon the deprivation of their lambs, farmers seem, in practice at least, to dis-
agree. The method of milking them is
not now so generally pursued, and seems
to be discountenanced by many of our
most respectable farmers. The different
situations and nature of some farms, no
doubt, renders it less prejudicial to the
stock than on others; but upon all farms
I am rather inclined to think, that it is
not productive of any considerable sum.
After much toilsome drudgery, indeed a
great quantity of cheese may be obtained,
and these disposed of on very equal
terms; yet the remote consequences afford
a counterbalance, at any rate nearly so,
for the gain thus painfully acquired. Besides the outlay for wages and milking

utensils, the ewes are damaged in no trivial manner. Huddled together and driven to the same place twice every day, where they are accustomed to the harshest treatment, the injury which they must sustain is easily conceived. In consequence of which a deficiency both in point of the number and strength of their lambs unavoidably follows; together with the reduction which it causes in the weight of every fleece of wool. With such unhappy results it can scarcely be expected that any benefit will in the end accrue from this system, and in my opinion it may almost be ranked amongst the unprofitable methods that have been employed by farmers for amassing wealth.

As soon as the lambs are taken from their mothers, some people have been in the habit of sending them to a different farm for the space of six or eight weeks. This plan, not very generally followed, is, I believe upon the whole, rather a good one, and which it would be better

for many to adopt. Their own pastures,
during the interval occasioned by their
absence, acquire an abundant growth,
which their continuance on them must
have prevented. It is, no doubt, attended with expenses, but these are not great
when compared with future advantage.
There is sometimes difficulty in obtaining
proper pasturage for lambs; but it is
generally got at three-half-pence a week
for each lamb, and which for eight weeks
amounts to one shilling. This will not
appear of much importance if we take
into consideration the great benefit which
they cannot fail to derive from returning
to pastures well grown, and raising them
to a full condition for the winter season.
But it is not every kind of food by which
when away they will be profited; it should
bear some resemblance to that of which
their own pastures consist. For if it is
of a finer quality they will fall away, instead of improving on their return; and

if much inferior they will suffer too rapid /
a decline when away.

This plan, however, will on some farms
be unnecessary, while on others it will
be unprofitable. They are occasionally
to be found with an extent of heath in
the most remote parts, which is of little
other service than as it may be appropriated for lambs immediately after weaning. In these cases no pasture need be
taken in a different quarter, and there is
another in which it might be attended with
a serious loss. I allude to the farms on
which the disastrous disease called Braxy
or sickness is prevalent. It appears that
the growth which the pasture acquires
while the hogs are away, tends to en-
courage, or is rather the principal cause
of the disease. It is indeed better to
have a death by this than by poverty,
but, if left to itself, in a short while it
may be like a destructive blast spreading
desolation all around. Before proceeding to any great length, however, it may

receive an effectual check from certain sorts of food, to the application of which I refer the reader to the end of the volume.

Before the commencement of winter, and about the latter end of autumn, the next circumstance that demands attention is that of salving the sheep with a mixture of tar, butter and milk. This manner of covering them is more commonly known by the name of *smearing* them. In high districts it is a measure that is absolutely necessary for the good preservation of the sheep, as they immediately fall away and cease to thrive when it is neglected. The greatest benefit, perhaps, of which it is productive, is that it effectually destroys the vermin by which sheep are infested. It indeed lessens very much the value of wool, but without it sheep are unable in hilly regions to withstand the storm as it rages and is felt there, and is also the means of preserving a great deal of wool which would otherwise have been lost. The proportion of tar to butter is in different parts of the kingdom far from being alike; but according to the general manner of salving here, one pint of tar and *3lbs.* of butter, compounded with as much milk as will render it soft enough to endure being laid on without breaking, will be sufficient to go over a half score. Care should also be taken that the divisions be not far separated, as the vermin will then collect between them, and, besides other effects, will very likely scab the sheep.

Having arrived at the period when vegetation retires from the earth, our observations must now have relation to the treatment of stock during the hazardous season of winter. The mildness and serenity of some seasons, indeed, renders in a great measure unnecessary any assistance and provision from the careful hand of man, but such may be considered as deviations from the usual

N

course of nature. And even in those
that are unmarked by any rueful blast,
the feebleness of hogs require some sort
of compensation for the general sterility
and roughness of winter. We have already in a former page had occasion to
recommend to the practice of every
stock farmer the plan of storing four-
fifths of his turnips towards the middle
of November. The gleanings that are
left upon the field, and which are supposed to constitute a fifth, are only to
remain there on purpose of being eaten
by the hogs. Immediately after the
others have been deposited in the pits,
they should have the privilege of con-
suming the remainder, that they may
not beforehand lose a single ounce of
condition. In high situations, however,
they should never be confined too close-
ly on turnips, as instead of increase,
their bulk will be liable to diminution.
So very hurtful does this sometimes prove,
that I have put a very fine lot of hogs

upon a field of turnips at Martinmas,
keeping them there without the least in-
termission in hopes of raising them to
full condition, and have seen them, to my
great mortification, dropping off through
pure poverty in the beginning of Feb-
ruary. This was not owing to any
lack of food, but altogether in consequence of the barrenness and exposure of
the situation, which with hogs, when they
are bound down to turnips in an open
unsheltered field, will always be found
to destroy in some degree at least, the
good which they might otherwise derive.
In being afforded the gleanings, therefore, hogs should only be confined on
them during the former part of the day,
and drawn off each afternoon to their
pastures until they are again returned
the following morning. When the gleanings have been all consumed after this
manner, the hogs must continue to be
supplied from the stores. These should
be led to places of shelter in convenient

parts of their walks, that after having eaten them they may fall on to their natural pasture. Turnips should in this way be laid down to them at the rate of a double-cart load for every eight score. To consume these about four hours will be requisite, after which they may a little before noon return to their common food. Such treatment should hogs continue to receive, till at least the beginning of March, and longer if the backwardness of the weather render it necessary.

If hogs are furnished with turnips in such quantities as are above specified, and for such a duration, they will be nothing reduced before the revival of spring. The small sum which has been expended or which might have been gained but for the sake of obtaining these turnips is not to be brought in comparison with the advantage they yield. Besides the superior value to which they raise the hogs in other respects, they are a security against death by certain diseases, and at any

rate by poverty; and I also experiment-
ally know that by the advantage of
them every fleece will weigh one pound
heavier than it would have done without them, which has always been equal to
half the value of turnips consumed by
each sheep.

Hitherto we have been advancing upon
the supposition of the absence, (and our
observations have consequently had no reference towards averting the horrors) of a
storm. It is a hope which the farmer in
a high district ought ever to banish from
his mind, the hope of the winter passing
away without as much as can be called a
storm; at least it would be the utmost measure of folly for him, in expectation of
the fulfilment of that hope, to make no
provision for one. The chance runs high
against him, as past experience justifies
the contrary conclusion. A storm is what
may be looked for, more or less severe,
every season before the winter months
have elapsed; and as the greatest difficulty

by way of provision for stock, is then to
be encountered, preparation somewhat proportionate to the demands of a storm,
should also every year be renewed. And
we ought especially to beware of setting
too narrow limits to what we suppose the
continuance of the storm in the ensuing
winter may be, lest we thereby regulate the
extent of our provisions. Many of us have
witnessed storms,—not sudden overwhelm-
ing blasts, the loss occasioned by which,
it is in our power to avert, only by means
of shelter,—but storms, whose protraction,
much longer than we expected and prepared for, has scattered the arrows of death
amongst our flocks, and left the impress of
want and starvation upon others that remained. Such mistakes of low calculation,
have frequently produced the most mem-
orable effects ; and our preparations should
be made, not for what generally is, but
for what we have seen to be, the duration
of the storm. Our sheep pastures are
very commonly blocked up, first and last,

for the space of six weeks, but then they are also occasionally and entirely covered for double that space. So that it is better to run no risk by laying up for three months, than to run the hazard of losing lives by a more partial provision. How much will be needed in the continuance of so long a storm, it is easy to determine, the quantity which every sheep will consume being accurately known; and if the snow dissolve in a shorter time, what remains untouched can be reserved for another year.

It will be found the safer and more convenient plan, as the Hon. Captain Napier recommends, to have the hay beforehand laid up at each stell, wherever stells are erected for shelter. According to the construction of stells, which I have formerly described, the hay, as was also then mentioned, should be contained in the area of the stell. Perhaps it would also be advisable to store a few turnips in some of the stells belonging to the hog pasture, as a heavy storm might prevent for a while all kind of communication with the different parts of the farm, and the hogs thereby deprived of their accustomed supply of turnips; which loss would be felt more severely in a storm than at any other time.

If the sustenance of sheep is to be altogether dependant upon what is laid down to them, ewes to be kept in good condition, will eat every day at least 1½ *lb.* of hay. Nothing less than this can be allowed them, or they will immediately begin to fall away. Hogs with this food alone will be kept upon 1*lb.* each. But if they are to be regularly supplied with turnips, as unquestionably they should be, at the rate of a double-cart load to every eight score, which will afford half support, half the quantity of hay also will then suffice, or nearly 4*st.* to every eight score.

On such liberal allowance stock will never fall in condition. They will not present, on the restoration of their pasture, a wan and languid appearance, as if they had had to struggle with the hardest difficulties, and been supported on the meanest fare. They will not present that shattered and emaciated form, to which they could only have been reduced by beggary and starvation, and look as if they had been but newly emancipated from a scene of wretchedness and misery. With such repulsive exhibitions we have been but too often familiar, and if the treatment above described be punctually observed, they shall be familiar to us no more. Our stock at the departure of the storm, will not seem as if they had been long encompassed with barrenness and sterility, but with the freshness and condition of a happier season, they will appear as if they had been accustomed with the verdure of spring.

Before taking leave of the storm, I have somewhat to remark respecting the

o

manner of feeding sheep with hay. Hecks are undoubtedly the best means that can be made use of for that purpose, provided there are as many as to allow the whole flock of sheep to be eating at once.   But to eat by rotation is at least a dangerous, and often a very destructive plan.   Of the truth of this I have been furnished with many examples, out of which I select the following.

A most respectable proprietor in my neighbourhood having some anxiety to try his skill in farming, took some of his farms under his own management. With what success he conducted his farming operations, save only the unfavorable one which I am about to relate, it is no business of mine to speak. As he intended to approve himself an exemplary farmer, he provided every convenience, and amongst other things hecks, for containing hay, out of which during the storm the sheep were to eat by rotation. Accordingly a storm came and that of no short continuance. The hecks were regularly supplied with plenty of good hay, but came far short in accomplishing the end for which they were obtained. Many of the sheep that were more modest, or who did not wish to push themselves forward by force to get their meat, were most miserably supplied; and out of a thousand ewes not less than ten score absolutely perished. But the ruinous effects survived the storm, and out of the eight hundred that remained, upwards of the half were unable to nurture their lambs.

Such a dreadful example as this it may perhaps be difficult to find, of the evil attendant upon eating by rotation. I doubt not that there may be instances produced in which few or no deaths have been occasioned, but still the evil is not obviated. Sheep that derive their sustenance thus, are subject to great disadvantage, and cannot but be very irregular in their manner of getting it.

If there are not a sufficient number of hecks to let them all eat at once, or at least nearly so, the hay had far better be given them upon the ground, or upon the snow after it becomes hard. In this way they will be kept in far better condition than eating by rotation; for though by the latter method it may be possible to escape absolute starvation, yet by it sheep can hardly avoid being reduced in their condition.

It may further be worthy of remark, that the farmer should beware of changing the food which he gives his sheep in a storm from better to worse. It would prove productive of injury to them were they first plentifully nourished with turnips, and after these were finished to have their food changed entirely to hay. If these different crops are to be given them at all, they should either be given them in conjunction, or the hay consumed before the turnips are applied to.

These are the most important observations that I have been enabled to collect concerning the treatment of sheep.
We have now conducted them through
every season of the year, and have again
arrived at the place whence we set out.
During the course of the summer they
become dinmonts or gimmers, according
to their sex, upon the simple process of
clipping. If, by the assistance of every
thing which has been mentioned as being productive of advantage, the ravages
of storm and disease have only deprived us of one since their separation from
their mothers, to every two score, we
have been very fortunate; and if, in
more perilous seasons, one to every score,
we have notwithstanding done pretty
well.

I cannot conclude without cautioning
farmers to beware of overstocking their
pastures. For besides diminishing the
profits which would accrue from being
more partially stocked, it is also one of

the most hazardous, and generally most destructive plans, into which it is possible to fall. It has often been the means of sweeping away greater numbers, than the most protracted storm. It is indeed, the remote cause of some of the most desolating diseases, which have been known to prevail amongst our flocks. It abates both of the quantity and quality of wool, and in all is so injurious to the sheep, that, even were it not the fosterer of disease, it would be attended with a considerable decrease of profits to the farmer.

Shepherds should also particularly beware of driving their flocks to and fro. Sheep thrive best when allowed to graze undisturbed throughout their pastures. If they are collected together and posted about from one place to another, in order to be brought to where the shepherd has assigned as their breakfast and their dinner lares, they will never, in the world, rise to good condition. And he ought too to be cautious not to over-heat them ; as when they are violently heated it gives rise to a dangerous and the most epidemical disease to which sheep are liable.

As the remarks that have been made on the rearing of sheep are all as general as possible, and as their treatment during the winter forms a very important part, it might probably be of some practical utility to subjoin an exemplification of what has been stated concerning the provision necessary for stock, and the proper manner of distributing it, by the example of a particular farm. To make this of a respectable extent, we may suppose it capable of supporting 1000 ewes and 17 score of hogs ; as this

number of hogs will be requisite to furnish about 16 score of good gimmers, to
fill up the place of as many draft ewes.
We may also allow 70 acres of land to
be kept in tillage, and this to be divided
into four bricks, producing in regular
rotation, turnips, corn and hay. By deducting two acres for potatoes for house
and servants, there will of course be raised annually 15½ acres of turnips. Taking an average crop, every acre will produce about 30 cart loads. Twenty-four
of the best from each of these must be
stored at Martinmas, or perhaps before
that period, constituting in all 372 cart
loads. This will all be performed in the
manner formerly described. The hogs
will then be put upon the fifth part that
remains in the field. In them they will
find an abundant supply for six weeks, by
taking them off to their pasture during
each afternoon and night. These being
all consumed, two cart loads should be
led to them every day from the stores, to

the most convenient places of shelter in their pasture for at least two months following. By this time 120 cart loads will have been finished, supposing them to have been laid down fair weather and foul. On this allowance the farmer will be enabled to rear a very good lot of hogs, and at the cost of about seven acres of turnips; which in a high district cannot be reckoned worth more than three guineas per acre. This comes to something less than 16*d.* for each hog, which over and above being a preventative from many diseases and poverty, is half regained by the additional pound of wool, which I have always found occasioned by turnips.

There still remains in store 250 cart loads of turnips, and the produce of 17½ acres of hay; which by allowing 120*st.* per acre, a wide enough calculation for a high country, is in all 2100*st.* In mild winters this will be little needed, and is mostly requisite for a

P

storm. In the event of a storm the
1000 ewes will require about 68*st.*
each day (*22lbs.* to the stone.) In
addition also to their two cart loads of
turnips, the hogs will also consume about
8*st.* per day. The hay at this rate will
serve for a month, which indeed is
much shorter than many storms. A
much larger provision should, therefore,
be laid up at first, and if the above
quantity of hay be annually raised, it
will prove sufficient, as the average continuance of storms will not be more than
a month, and what is saved one year
can be reserved for another.

If the 252 cart loads of turnips can
be saved, as in all probability they will,
they must be given to the ewes in the
spring. The 16 score of gimmers
should always be separated from the hirsels about a month previous to lambing
time, and three cart loads given them
every day. At the end of the month
84 will have been done. The whole

number of ewes and gimmers should
then receive the remaining 168 cart
loads, at the rate of six per day. I
leave it entirely to experience to justify the reasonableness of these measures.

CHAPTER V.

ON FEEDING CHEVIOT AND BLACK-FACED SHEEP.

———

As it is only my intention to offer a few cursory remarks on the mode of feeding sheep by turnips, our observations must necessarily be confined within very narrow limits. I wish not to say any thing of the other methods that are sometimes made use of to fatten sheep, as they are not of such common practice and are not so applicable to stock belonging to a high country. There are perhaps few articles that would tend more directly and more effectually to accomplish that end than salt given them at proper times and in proper quantities, along with their other food. The fattening effects of salt may be seen by turning to other parts of the treatise, and I leave it to others to make the experiment.

Those stock farmers who make a practice of feeding the sheep which they rear, must either themselves possess land in a more fertile district, or purchase turnips from those who do possess such lands. In stock farms of upland situations there are no turnips, at least there should be none, adequate to feed for the butcher even a small number of sheep. Land of good quality indeed, may be found in some of them, on which it may be possible to raise more turnips than are absolutely necessary for the maintenance of the ewes and hogs, but these if given at all to feed sheep, can only be given during the former part of the winter, as the rigorous colds that are customary in such exposed situations are very injurious, especially to sheep preparing for the butcher, and in a great measure destroy the nourishing effects of the turnips.   Whenever in such cases a few acres can be spared, they should either be appropriated for bettering a little, in the beginning of the season, the condition of sheep intended for the butcher or led home for fattening oxen.

Sheep for fat ought to be put upon turnips about the middle of October, or rather at an earlier period, if they are then making no advancement in condition.   And when they are granted them, one lot should never exceed in number 20 score.

Whatever quantities of turnips may be obtained in a high situation, and at however cheap a rate, unless they possess every advantage of shelter, the sheep should be removed to the low country, if not in October, at longest in the beginning of December.   The field to which they are taken should by all means be dry, for if it be of a wet nature it will prove of a hurtful tendency. Its size should also be proportionate to the number of sheep by which it is to be eaten, as it is undoubtedly a disadvantage for them to be changed from one field to another.

Care should also be taken not to give them too much scope over the field amongst the turnips, for in this case they will injure them very much, and the latter will become before they are consumed dirty and not fresh. One brick should occupy no more space than will suffice for food during one week. And in being afforded a new brick, their liberty should also be extended over the ground they have already broken. The gleanings upon that brick to which they were last confined should not be picked until the sheep are let upon another one, and they will then naturally fall back and eat up the shells which they had left behind.

Sheep upon turnips though not so

liable to sustain hurt from a storm as those that have only pastures, are not yet altogether exempt from any evil being occasioned by that cause. In a storm of long continuance, turnips that remain in the field get so excessively hard as to render them in a great measure incapable of being eaten by the sheep, at least not near so readily as they are accustomed to be in fresh weather. Sheep, in a long unmitigated frost, may thus have many a hungry day; many not so plentifully supplied at any rate, as in the absence of frost. To avert any damage being incurred in this manner, a few turnips should be stored before any storm has arrived, and laid down to them when they cannot derive so much benefit from those in the field. If some measure of this kind be not adopted, sheep will never feed fast in a storm.

Hay is an advantage to sheep feeding upon turnips too considerable to be

overlooked.   If it can be found good
they will relish, it and eat a little of it
every day, being a change.   It should
be put into hecks, and these placed at
convenient distances throughout the field.
Natural well win bay will answer the
purpose as well, probably better than
any other, and can be furnished at a
cheaper rate.   Perhaps it might be advisable to sprinkle this hay with a little
dissolved salt; the sheep might then
consume it with an increased fondness,
and undoubtedly it would yield them
more nourishment.

I have invariably found that sheep feed
faster upon turnips as they grow, than
when pulled and led into a different field.
They will improve more speedily in condition by eating them in the former way,
than if they were to be brought to them
even in a grass field.   When laid down
pulled they lie in such a loose and un-
firm state that by their rolling about
the sheep can never obtain a substantial

Q

hold of them, and more especially when they are hardened by frost. Sheep also acquire an unsettled habit by continually running after the cart by which the turnips are conveyed to them, which is in some degree prejudicial to them.

There appear to be few other remarks on this subject worthy of notice. With the manner of feeding sheep small difference seems to prevail, and the usual way is pretty fully understood. Wethers of the Black-faced breed are seldom raised above 13*lbs.* or 14*lbs.* per quarter, and are more commonly below that weight than above it. Those of the Cheviot generally feed to the weight of 15*lbs.* or 16*lbs.* per quarter, though I have very frequently brought them to a pound or two more. Once, indeed, I had them of an amazing weight, and what they never equalled with me either before or since. Five wethers I retained to exhibit for the premium at Coldstream, in the year 1818, for which I obtained as many guineas. They had been fed upon turnips alone during winter, and were sold to a Berwick butcher at the high price of £4 10s. each, which together with the premium amounts in all to £5 11s. for every wether. They averaged 30*lbs.* per quarter upon being killed, and one of them weighed no less than 32½*lbs.* Very few instances of young Cheviot wethers being raised to such an astonishing weight, and being sold at such a high rate, can perhaps be produced.

CHAPTER VI.

PRINCIPAL CAUSE, AND DESCRIPTION,

OF THE ROT.

———

𝔄 GREAT many diseases have been enumerated by writers as incident to sheep. Many of them, however, are either altogether foreign to Lammermuir, or appear so slightly that they are seldom dreaded. But though few, there are some that have appeared with the utmost vigour and the most unabated force, and with which we have unhappily been but too much familiarized. To the most formidable of these (which are chiefly three, the Braxy or sickness, the Sturdy or water in the head, and the Rot) and that the most destructive disease

that perhaps ever desolated our flocks,
I mean at present to confine my observations. As this, namely the Rot,
has unfortunately come more immediately
under my own inspection, in treating of
it, I shall be guided by facts that
mostly occurred on my own farm, or
should I at times step beyond this
boundary, the materials shall be collected from among my neighbours or
other genuine sources, and none shall
be produced that are not perfectly legiti-
mate, and can be well authenticated.

The fatal experience of many will
justify me in saying, that there is no
other disease more worthy of the most
scrutinizing research and diligent investigation, than the one concerning which
our remarks are now to be directed.
There are none whose terrifying ravages
have more extensively and more universally prevailed, and none which has tended more readily to disburden farmers
of the profits that they had laboured to

accumulate. It is to be dreaded as a pestilence and every means, which it is in our power to employ, must be used either to prevent it altogether, or if possible, when these are neglected or are ineffectual, to cure it when it makes its appearance.

Much has been said and written on this important subject, and not entirely without effect; but there is still a wide field left for the exertions of those who by unwished for experience have known too well its afflicting consequences. Amongst this class I am sorry to state that my loss will place me if not altogether, at least very near the head of the list.

In 1810 my stock consisted of 2000 ewes, hogs and dinmonts, out of which I lost, by rot, during the winter and spring following above 800.

In 1817 I lost 900 of the same complaint, and as a number of them were ewes, I found a deficiency of 400 lambs, at the time of weaning. Many years preceding the above I had severe losses, though never to such a ruinous extent. I have, therefore, mentioned these years as being the most destructive of any in all my experience.

After having endured heavy strokes like these, it is hardly necessary to observe that I was led to try many things to check the wild impetuosity of its career, and to prevent the recurrence of such a calamity. And I am now happy in announcing that I have been enabled to bring forward a cure, which if rightly acted upon will prove of essential service to breeders in general. In stating this I lay claim to no original discovery; I have only brought here more fully into notice, and ascertained by actual experiment what has frequently been mentioned before, but never fully, in Britian at least, brought into action.

Before entering more at large on this subject, it will be proper first to enquire

into the origin of the Rot, ere we point
out the means of preventing, and prescribe the cure by which it will be
quickly made to disappear.

Concerning the causes that tend to
produce the rot, a diversity of opinion
seems to prevail. Mr Hogg, the Etterick shepherd, in his treatise upon the
diseases incident to sheep, enumerates
several causes which are mentioned by
others as being conducive, and which he
either combats with the view of establish-
ing his own theory, or resolves them into
it. With a sweeping conclusion he at
last ascribes it to the want of food and
shelter, and " holds as an incontrovert-
ible fact, that *a sudden fall in condition*
is the sole cause of the rot." With all
deference, indeed, to such good authority, I presume I shall be able by a plain
statement of facts to make it not only
controvertible but also to disprove it, at
least as being the sole cause. Hunger
and cold, no doubt, are the parents of

many dreadful calamities among the human species, and I would be inclined to allow them their full preponderancy in the diseases of other animals, but as many of our own species are liable to diseases and death who are altogether exempt from both, so among sheep I have known many hundreds die of rot, where these causes could never be brought, in the most remote manner, to have any share in the account.

In 1816 and 17, the Lammermuir farmers, and I may say breeders in general, suffered in many respects from the severity of the seasons, and I believe the latter was the most general rot ever known in Britain. Now if we can be able to lay hold of any circumstance peculiar to these seasons, it may lead to a reasonable conclusion as to the cause of this complaint. And surely every one concerned in the management of stock at that time does not need to be told

R

that both seasons were wet even to a proverb.

The year 1816 besides being wet was also extremely cold, and many store farmers dreaded the rot would be the consequence, but at that time were happily disappointed. The reason undoubtedly was, that though the season was wet, it was below the average temperature of seasons which fortunately prevented any after-growth of grass. Had the rot then been prevalent it would have given force to Mr Hogg's theory, but the contrary being the case it undoubtedly invalidates the strength of his reasoning.

The year 1817 was again very wet, rather more so than the one preceding, but then the average temperature of the season was several degrees higher than the other, which produced a very abundant growth of grass in the months of September and October, and the ultimate consequence of which was that one of

the greatest fatalities by rot followed to
which the memory of man bears evidence. Now I hold it almost amounting
to certainty that the after-growth of
grass in the months of September and
October is the great or chief cause of
the rot. No doubt many things may co-
operate as predisposing previous to this
exciting cause, and it may even be said
that the cold and wet of 1816 may have
laid the foundation of the fatal rot of
1817. This I shall not positively deny,
though, had the seeds of the complaint
then been scattered undoubtedly some
symptoms would have appeared, and
though my experience rather goes to
prove the contrary, and as I am more
inclined to abide by plain matters of
fact, than enter on any visionary theory,
I shall content myself with merely stating the grounds of my dissent from this
opinion.

All my sales made in 1816, were perfectly sound, and in the year following

down to the month of August, not the least symptom was present that could in the smallest degree justify the suspicion of any complaint being among them. In June, 1817, I sold a lot of about 1000 hogs and dinmonts to one gentleman in the county of Roxburgh, all of which gave the greatest satisfaction. They were kept by the same gentleman for two years, and afterwards sold in the finest condition to the butcher. This was well for both parties, but the sales that I made in October were all tainted, and from that time they consisted more of skins than carcases.

Here then the facts bear me out in saying that in 1817 no rot had taken place among my stock in the month of August, and the whole calamity that followed must have taken place subsequent to that period. Had any latent seeds of the disease been among them, the sales that I made in August must have turned out as bad to the purchaser as those that were retained did to myself, which was not the case, and which clearly demonstrates that the cause had been on my own farm; of this I entertain not the smallest doubt, and after the most minute investigation can attribute it to nothing but an unusual luxuriant growth of grass occasioned by the mild soft weather during the months of September and October, more especially during the first.

This tender but destructive sort of grass is also sometimes produced by other means, such as by horse and cattle dung dropped during the preceding months of summer. And here by the way I would strongly condemn the practice of allowing sheep and cattle to pasture promiscuously together; for in many cases it may be cause of rot where none would have appeared.

The rot may also be occasioned by the succulent herbage that grows upon flooded water sides, after shaken corn, and recently improven wet-bottomed moor producing a soft and rapid growth. This Mr Hogg opposes upon the faith of a correspondent who considers their eating this as the consequence not the cause of the rot.

It is a curious and important fact that the fluke-worms are found in the livers of all rotten sheep, and I have no doubt of these insects being the immediate cause of death; but how they come there has never yet been properly accounted for. We cannot suppose that they form part of the original mechanism of the animal, inherent in its constitution and only called into existence by certain fortuitous circumstances; this I think would be venturing too far upon hypothetical ground. It would be more consonant with the operations of nature to suppose the eggs of these animalculæ taken in with the food and carried along the alimentary canal till they are again taken up (in conjunction with the chyle) by the lacteals and conveyed through the mesentery into the thoracic duct, whence they are sent into and mixed with the blood. They may thus be transmitted by the circulating fluid through its various conduits, till they arrive at the liver. To this viscus the blood is sent in great quantities from the spleen, mesentery and stomach; the vessels from each of these uniting form one large vein which enters the liver, and thence divides into innumerable branches, which at their very minute ends form an immense number of vessels arranged like the hairs of a pencil brush, and hence called (in the human subject) *penicilli*. These penicilli constitute the glandular fabric and bulk of the liver. Here the capillary vessels obstructing their further progress, and affording a proper situation for hatching, the worms may be produced and bring on that fatal disease called rot.

The above is only brought forward as a probable conjecture, and has been mentioned by others in a somewhat similar

manner.   Though the passage of these
eggs into the liver seems to be beset with
difficulties, and apparently hardly possible
for them to escape without their being
injured, yet this appears to me to be the
most seemly way of accounting for their
getting there at all.   After having reached the liver it is no improbable thing for
the eggs to be hatched there, for we have
the authority of the great naturalist Spal-
lanzani, who says, " If vegetable seeds ger-
minate without exception in confined air,
what are we to think of animal semina or
the eggs of insects, which according to
Boerhave, and the general opinion of phi-
losophers, should remain sterile, even when
the operation of circumstances the most
favourable to their production occurs?
Here I thought it better to consult nature
than to trust to the sentiments of others. I
therefore made experiments on many eggs :
on those of beetles, flies, flesh flies, noc-
turnal and diurnal butterflies, worms and
others, and scrupulously observed what

happened to each kind. I foresee the reader's anxiety to learn the result of these experiments; and in two words his curiosity may be satisfied, by learning that the whole different species were produced equally in confined as in open air."

But in whatever way these worms are produced the fact is unquestionable that they are always swarming in the liver of every rotten sheep; and that in proportion as a sheep is far gone in the disease, the more numerous do they become, most certainly the two have some connection with one another, and that no small one, but whether they are the cause or the consequence of the rot, remains yet to be determined. As Mr Hogg says " it is a curious circumstance, that of all other diseases of sheep, the greatest variety of opinion prevails with respect to the real cause of this; and among such a number, it may reasonably be expected that it is very difficult to alight on the right one." This great diversity of opinion he has I think accounted

for in another part of his treatise; " That
the diseases of sheep are numerous and
complex is too well known; yet from their
extraordinary fewness on some farms compared with others of the same nature, and
on the same farms under a different management, I am often tempted to conclude,
that they are naturally as free of them
as the hawk or raven; and were I able
to define the various parts of the animal
frame, their connection with one another,
with the influences of climate and regimen upon each of them, I have no doubt
but I should make it appear that the whole
of the diseases to which this useful animal
is subjected, might be traced to have originated in accidents proceeding from improper usage or inattention in their keepers
or managers. Soils and seasons have their
influences, and that to a degree so extensive, as that they will never be entirely bettered; yet still they may in a
great measure be guarded against." The
difference of soils, seasons and manage-

ment, thus elegantly stated by the Etterick shepherd, accounts fully for the difference of opinions concerning the diseases to which sheep are liable, and he
thus elegantly concludes; " For my part
I anticipate with exultation, the ap-
proaching happy era in the history of
farming, when the *Rot* and *Braxy*, which,
in their respective districts, have raged
like a pestilence among the woolly tribes,
and buried the hopes of the husbandman with his bleating flocks, shall be as
much eradicated as the small-pox is, at this
day, among the human race. For to what
an extent has their rigour been abated,
even in our remembrance? On many
farms, where they cut off annually about
a sixth of the stock, their baneful
influences are now scarcely felt." And
I hope it will not be deemed presump-
tion in me when I add, that I trust
the happy era on the contemplation of
the arrival of which Mr Hogg dwelt
with increasing pleasure, respecting the

rot at least, only waits for opportunity and proper application.

But to return to the cause of the rot. The one assigned by Mr Hogg is very far, I regret to state, from according and is actually at variance with my experience. In no case that has hitherto come under my observation has a *sudden fall in condition*, in the smallest degree contributed to bring on that mortal ravager; nay, in many cases with which I have been most intimately acquainted, it could neither be traced, with the strictest scrutiny, to this source, nor did this follow even as the consequence of the disease.

Once, indeed, that opinion had also gained my assent, and in conviction of its truth I acted upon it for many years. It is undoubtedly the farmer's interest to have his flocks at all times in the best possible condition, as in that state they can always be turned to the best account; but, as far as I can rely upon my experience, it can never form a barrier against the rot.   In the lower parts of Berwickshire, where they were treated with the most scrupulous attention, and where food and shelter abounded, I have known many scores of sheep fall victims to this disease.   I am not at liberty to mention names : but one famous instance I would bring forward, of one of the most distinguished breeders in this fertile district, whose sheep possessed every mark of the prevalence of the disease, and which, though never known to have been reduced through the whole term of their lives, were yet dying of the rot, in the greatest numbers and in the highest condition.   The proprietor's anxious hope that the ravages of the disease would quickly be at an end, and his unwillingness to part with his stock, which notwithstanding were in the finest order, made him keep them on much longer than he should have done : till at last seeing his hopes defeated and his sheep rapidly disappearing without any return, he was obliged to kill them by scores and cart them over the country to be disposed of for what they would bring.

My own affairs also yielded abundant confirmation of the insufficiency of food and shelter. In 1810, I put a fine lot of dinmonts upon turnips before the Martinmas, though all in very favourable condition, as I was rather beginning to suspect that they were affected; and under the idea that meat and shelter would provide against every exigency I sent them from my own farm, to a fine dry, well sheltered situation in the middle part of Berwickshire, where I expended no less than £100 upon turnips; but before the month of March there were few of them remaining in their skins, and I did not realise as much as defrayed the expenses laid out upon the turnips.

In the month of October the same

year I bought a lot of wethers in fine
condition, from land of a good sound
bottom, where the rot was altogether a
stranger.   They came on my farm about
the middle of the month, and in a short
time I observed they were all affected.
The stock on the farm whence they were
taken, continued all sound, so that the
complaint must have originated with
myself: not in any sudden fall of condition, however, as I conceive, for none
was observable, but in the soft luxuri-
ancy of this part of the season.   But
though the whole were evidently tainted,
my loss in this case was unexpectedly
small.   I put them on good grass early
in the spring, and sold them to a butcher in the month of August in very
tolerable condition.   This case may seem
to favour Mr Hogg's hypothesis a little,
inasmuch as sheep in high condition
and health will be able to hold out
longer by mere strength of stamina.
But I would rather be inclined to con-

sider it in a different light, and as one of
the most forcible examples that could be
adduced in opposition to his theory. For
as in my opinion frost always puts a check
to the extension of the disease, preventing the further growth of soft grass,
the only circumstance in the above case
that could tend to abate its destructive
consequences, was that the infection was
not caught till the middle of October.
In so late a period of the season
there could be no continuance of soft
rainy weather without frost, so that the
seeds of the distemper would hardly be
engendered until they were again deprived of nourishment. For had it happened
at an earlier period, and during the
month of September, which is the most
dangerous month, and when frost is less
to be expected, the whole flock would
undoubtedly have been consumed, as my
experience since leads me to conclude.
From that time I have invariably found
that when there this a continuation of

weather favourable for producing it in September, and when it is then fairly begun, it bears along with it every thing that opposes its progress and melts down the most robust constitution.

In 1817, my stock was in good condition at the end of August, and the first that died of the rot was in a high state of condition considered as hillstock. And in the lower part of the county the same year, a great many farmers sold their whole stock to the butcher, when they saw them effected. I knew several of them who sold their ewes at 30s. and upwards, so that we cannot suppose there had been any great or *sudden fall of condition*, in their case.

One other instance more shall, for the present, suffice. A friend of mine who had a pretty extensive concern, though rather in a high situation, but the land all improved and well sheltered. His holding stock consisted generally of 300

T

ewes, and 120 ewe-hogs. He began to suspect their being unsound early in the season. Under this impression, but not till after he experienced considerable loss, he began driving them to the Edinburgh market. Many of them dropped down dead upon the road; what survived of them were sold, the ewes at from 26s. to 27s., the hogs at 18s. and 19s. Here again the want of condition seems to have had nothing to do in the matter.

To carry on this train any longer would be both tedious and unnecessary. To any candid inquirer it will surely have appeared that, in whatever the rot may originate, a sudden fall of condition has no share. And if any person can come forward and prove that it is *not caused* by an over-abundant growth of grass about the middle and the latter end of Autumn, either occasioned by the state of the weather or any of the other circumstances formerly specified, I shall

freely grant that, with our present knowledge, the true cause still lies hid in
the dark recesses of nature.

It remains now for us to give a description of the most evident symptoms
by which the rot on its appearance may
be easily discerned.   This has already been
done by Mr Hogg in a more distinct and
comprehensive manner than it was possible
for me to have done.   And I trust I will
therefore be excused for inserting here the
most important part of his treatise concerning the symptoms of the rot, and also
a little from the appearances on dissection.

"The first symptoms of this malady
among the flocks should be guarded a-
gainst with the utmost care and perseverance, which are as follows;—When a
severe storm of snow covers the ground,
and locks up the herbage, so that they
cannot attain nearly a sufficient quantity
of food for some length of time; or when
the weather is so boisterous that they cannot stir abroad to shift for food, or when

they receive any bad usage; if subsequent to any of these, or indeed on whatever occasion, a lethargy prevails among them; if they grow dull and careless of feeding, the rot is certain to make its appearance by-and-by; if this lethargy be general, the rot will also be general; if it prevails only with certain individuals, these are they which the rot will affect," or as I should say are already affected.

"The next symptom that is discernible after this lethargy is in the shape; the belly being shrunk, and clinged up for some time; they then fall to their meat with great voracity, and as long as their bellies continue light, they are not quite fallen a prey to the disorder. After this clungness, the belly falls down, and the flanks fall in, which is a worse symptom, as is natural to suppose, the disease being then a stage farther advanced."

"When a shepherd, or farmer, is endeavouring to discover such as are unsound in a fold, let him feel the heck, or

small of the back; and if the ewe be
firm there, and the skin refuse to slide
on the flesh, it is a good sign, and if she
be not too old, is safe to keep. Lean.
ness on the brisket, or ribs, is not so bad
an omen of the rot; but a lean back is
ever dangerous where the rot prevails, or
is suspected. When he lays his hand
first upon the sheep's back, or ribs, let
him do it very softly, and press it still
harder by degrees, and if he feel a slight
crackling, as if there were small dry
bladders betwixt the skin and flesh, that
sheep will invariably turn out rotten,
and is indeed so far gone, that she
is past redemption to all intents and
purposes;" unless, he might have said,
some restorative of great efficacy be immediately applied.

"Recourse must next be had to the
eye, which is an invariable rule to judge
of the state of the liver, and fountain
of life. Let the corner of the eye, next
to the nose, be turned out with the thumbs

pressing gently on each side, and if it be streaked with beautiful red veins branching to and fro, the sheep is safe and sound: the redder that the eye is the better; but as grass-fed sheep's eyes are never red, if they are free of a watery gilt, not too thick, and above all streaked with red veins, there is no fear: But, on the contrary, if the eye is yellowish, clear with water, and no red veins branching through it, the sheep is certainly unsound."

As to the poke which they acquire below their chops, it is certainly a sign of the prevalence of watery fluids over the vitals at that present time; but it is not a certain sign that the animal is lost; for, on the contrary, a very lean rotten sheep is most apt to overcome the rot; and such sheep as, by mere oppression, are rotten on hard heathery lands, very generally have the poke: yet these will frequently, in a great measure, get the better of it; and all ewes that are visibly affected by it, are better with lambs sucking on them than eild, for if they are eild, they are at-tacked by a lingering dysentery, which gradually brings them to their end.

" The next thing whereby to judge is the mouth ; for if the tongue be red and clean, it is a good sign : but the teeth must also be minutely judged ; because if they are kept a year or two over old, they are apt to decay before next year's draft ewes go away. Now, the age of a sheep is very easily known by its teeth; for in its second year it hath two broad teeth in the middle; when in its third year it hath four broad teeth; and, while rising its fourth year, it hath six broad teeth ; next year its teeth are all cast, and consequently are all of those called broad teeth ; and when it is five years old, and rising six, they grow as narrow at the top as at the root, while, as before, each tooth spread at the top." But much of the trouble requisite to examine the teeth of sheep,

In order to determine their age, may be spared by ear-marking them.* It is dangerous and unprofitable in more respects than one to keep ewes after they are five years old.

"When a sheep is killed during the early · stages of the disease, about . the time when the flanks fall in, the fluke-worms are only to be found in the ducts of the liver, but often in great numbers. The liver itself is by this time swelled a full third larger than its natural size, and seems to have undergone a considerable inflammation; its coat is thick, and of an opaque colour, resembling a pale clouded flint, or pebble. Nothing can be seen to ail the lungs. The tallow that covers the bowels and kidneys is loose and flabby; and looks as if part of it were melted, or its surface greased over with melted butter. One half or at least one third of this tallow

---

* See Page 89.

will not melt by any force of fire, and
such of it as is refined, and made into
candles, wastes and runs excessively."

"When flaying them, the fell is so
loose upon the back, that it will not se-
parate from the skin; on all other parts
of the body, the skin comes easily off;
appearing as having, in a great measure,
been separated from it before. The skin
is tenderer than that of a healthy sheep;
and the wool, though not yet loose, comes
more easily off. The fell about the short
ribs, and ribs, which, in a healthy sheep,
is a bright red, in them is pale, and bor-
dered with a tincture of yellow. Their
mutton when cold, does not grow stiff;
and when boiled does not grow tender,
but shrinks amazingly. Their blood is
thin, frothy, and light coloured; and
their heart, though perfectly sound, appears to be somewhat larger. As the
disease advances, all the other symptoms
continue to increase save this; for the

U

heart is often larger in the first stages of
the disease than in the last."

"When they die of the rot, the skin
and entrails are perfectly rotten, and the
whole body of a dirty pale colour. The
carcase has no peculiar smell; and although
when killed in the last stages of the disease, a large quantity of frothy blood
frequently flows; yet when they die of
the rot, very little blood is to be found
in any part of them. When the poke
below the jaws, is opened, the skin is
thick, and a congealed substance within
it; and within that a quantity of clear
water. The liver is the most horrible
mass of corruption and disease that can be
conceived. It is increased to two or three
times the size and weight of that in a
sound sheep; all its ducts and vessels
are crammed with flukes, and many over-
grown ones often on its surface: it is
half covered with hard white lumps of
various sizes, which, when cut, have a
grisly appearance; part of these are like-

wise mixed through all its interior, as are
also long layers of sand, so that it is often
hard to cut. The spaces, where none of
these intervene, have quite lost their con-
sistency, and are flaccid and gory. When
it is boiled in water, it grows perfectly
friable, and breaks in pieces of itself, while
that of a healthy sheep grows firm and
solid."

CHAPTER VII.

THE BEST MEANS OF PREVENTING AND OF CURING THE ROT.

———

IT is evident that in treating of the cause of this complaint, every thing that was then mentioned as in any way contributing to bring it on, must be avoided in our endeavours to prevent it. This part of the subject, therefore, has been partly anticipated. But the great remote cause which we have all along pointed out as being the most universal is a great fall of rain in Autumn, accompanied by a mild softness, which together produce a luxuriant growth of grass. The great means for preventing the rot, therefore, must be to destroy the immediate effects of this weather.

There has perhaps been no greater improvement in modern husbandry than surface draining upon soft and wet lands; and still more, if any part of the pasture is liable to be flooded by the overflowing of small burns and streams of water, after heavy falls of rain. In these cases the surface drains collect the water that would otherwise cover a large space, and carry it off in a few hours, where it must have stood for many days or weeks in a stagnant state, till it had either been absorbed by the ground or evaporated by the sun. By these surface drains the land admits of a very high improvement; the grass of which it is afterwards productive is of better quality, and has no tendency to cause diseases, that would in all likelihood be the consequence of the land remaining in its natural state. Such bogs, however, as produce deer-hair and ling, should not be drained; as the wetter the ground is, the more do these herbs grow, and of both the sheep are particularly fond. They are also very wholesome.*

Care should be taken that drains be properly laid out, so as to answer the purpose intended. It would undoubtedly be proper to take a general survey of the piece of ground to be drained as to its bearings and declivity; after which the course of the drains may be marked out, taking care at the same time, to avoid as much as possible all quick turns, as at these places the water is ready to run over, and defeat the object intended. The drains themselves should be about two feet wide at the top, and fourteen inches deep, with a gentle slope from top to bottom. Such drains are generally done in this neighbourhood at one penny per rood, and

---

* See Mr Hogg's treatise upon the means of preventing the rot. See also the Honourable Capt. Napier's treatise on Store farming.

even at that small price the workmen
make very good wages. The sods are
taken out the whole depth at once, with
an instrument for the purpose, and should
be laid down on the side of the drain
with the grassy side uppermost, as the
grass in that case will continue to grow,
whereas if they are turned upside down,
they will produce no grass for some
years, and there is consequently a loss
of twice the breadth of the drain. They
would also require to be occasionally
cleaned out, more or less frequently as the
particular nature of the ground may
render necessary.

I have been the more minute on
this head, as I consider it one of
the most effectual methods that can be
devised to meliorate coarse and soft pasture land, and as tending to improve
the stock both in constitution and condition. With this improvement too, a
greater quantity of stock may, in most
instances, be kept.

It must be evident to every one that where the system of surface draining is adopted, no cattle should be allowed to pasture; for besides being hurtful to the stock, and if numerous may themselves be the cause of rot, they very soon tread down all the sides of drains, and in a great measure render them useless. The shepherd's cow is undoubtedly one too many, but in many cases that cannot be avoided.

It may not be improper to take notice here of Mr Hogg's hypothesis. "Some tell me," says he, "that there are two kinds of rot: the black rot, and the hunger rot, the one occasioned by foul food, and the other by getting much too little food of any kind." The black rot he treats as altogether chimerical, and asserts that hunger and cold are the only causes of rot. If this be really correct, the means of prevention are obvious and need not be mentioned. Whether or not, indeed, hunger and cold are in any measure conducive to the rot, they ought to be especially provided against by every store farmer. But I again repeat that I cannot acquiesce in the opinion of these being the causes of rot; for throughout the whole range of my experience I have never found a single instance that could in the least tend to countenance the suspicion of these being the ultimate and principal causes. I have lost from time to time a great number of hogs by poverty, and I could certainly trace their death to "want of meat and shelter" alone, but there were none of those diagnostic symptoms apparent, which indicate the complaint called rot. Cold and frost are always severe on hunger-stricken hogs, but I have uniformly found, that frost prevented the rot, and that if the disease had not taken place previous to the arrival of frost, it never followed that kind of weather. Change-

x

able weather too, which is always hurtful to the condition of sheep, when it
partakes of a considerable proportion
of frost, is seldom, if ever, known to
produce it.

The state of the weather is, no doubt,
altogether beyond our controul, and all
the efforts we are capable of employing
to prevent its consequences, must in
many cases be feeble and insufficient.
Drainage and shelter however cannot
be too highly recommended, whether
considered merely as antidotes against
the evils that result from the extremes
of weather, or simply as preventatives
of the rot. For though cold may not of
itself be the source of the disease, it
may yet rapidly increase it, after it has
commenced, and greatly accelerate its
baleful effects.

These measures, however, effectual as
one of them at least has been, I would
only consider as auxiliary means for preventing the rot. It may happen, and it

actually has been found, that where these precautions have been taken, the rot, though unquestionably very much diminished in severity, does not cease occasionally to appear. And if further measures be not adopted, it will never cease altogether to appear on many farms, in those seasons which have the greatest tendency to give rise to it. There is another preventative which I would rank still higher than any other that has yet been employed; which, if properly put into execution, will, I trust, totally annihilate the rot, and on the efficacy of which I shall now enlarge.

The use of salt has been long known as contributing most essentially to the health and condition of quadrupeds in general. In many instances, where the industry of man could have no concern, nature seems to point out the utility of it, and in following her dictates we are never led astray from the true path of knowledge. In the interior of America,

x 2

where the undomesticated animals roam
at large, without being at all restricted
by the controlling hand of man, they re-
sort for hundreds of miles from all quarters, to those mountain sides where the
rocks contain salt, in order to lick the
saline efflorescence which is extracted
from them by the influence of the sun.
The avidity with which they lick this,
and the nourishment which it affords, is
easily conceived from the exertions which
they make to obtain it; and the same
having been done for many centuries, by
thousands of animals every year, and
that in a wild uncultivated country, the
tracts leading to these rocks are so beat-
en for miles round, and the places them-
seves trodden, and even excavated, to
such a remarkable degree, that they are
noticed by travellers as land-marks in
exploring these extensive regions.

That sheep in their natural or wild state,
following their own instinct, have recourse
to such means for their preservation and

proper support, we have the authority of the learned Zoologist Dr Pallas, to whose reports on these subjects we have already had occasion to refer, and therefore, our remarks shall now be more confined. When treating of the *Argali* or wild sheep of Siberia, he says, " The argali purges itself in the spring with acrid plants of the anemonoide kind, till milder plants spring up, and shrubs begin to sprout, which, with alpine plants, constitute its usual food. It likewise frequents the salt marshes, which abound every where in Siberia; and even licks the salt efflorescence that rises on the ground, a regimen that fattens them up very quickly, and fully restores the health, vigour, and flesh they had lost during winter, and during the purging course, so wonderfully dictated to the sheep species, together with the restorative, by the Almighty, whether in a wild or tame state if left to roam at large where the necessary plants are to be found." Here we have the very highest testimony, and the more to be depended upon, as he had no favourite theory to support, but merely relates the facts as they were witnessed by himself.

There are also many places where people, seeing the manifest advantage and nourishing effects of salt, use means to supply their flocks with it, where nature has not furnished it in their pastures. This practise, I believe, is no where more generally and more carefully attended to, than in Spain; and of the manner in which they there give it, we have the following account in a letter included in the Scots' Magazine for 1764. "The first thing the shepherd does when the flock returns from the south to their summer downs, is to give them as much salt as they will eat. Every owner allows his flock of 1000 sheep, 100 aroves or 25 quintals of salt; which the flock eats in about five months. They eat none in their jour-

hey, nor in their winter walk.—The
shepherd places 50 or 60 flat stones
at about five steps' distance from each
other, he strews salt upon each stone,
he leads the flock slowly through the
stones, and every sheep eats to his lik-
ing." Here follows a curious remark,
viz. "That the sheep in Spain never
eat any salt when feeding on lime-stone
land, whether it be on the grass of the
downs, or on the little plats of the corn
fields after hearvest-home." The writer
attempts to acount for this peculiarity in
rather an unsatisfactory manner, which
it is unnecessary to mention here. "The
shepherd nevertheless," he continues,
"must not suffer them to remain too
long without salt, and he therefore leads
them to a spot of argilaceous (clayey)
soil, and in a quarter of an hour's feeding they march to the stones and devour
the salt. If they meet a spot of the
mixt soil, which often happens, they eat
salt in proportion."

Thus in Spain salt is considered a most essential requisite to sheep, and we find that in some parts of America the same method, though after a different manner, is also practised. A letter from a gentleman in America, included in the above work, clearly points out the great benefit of salt, not to sheep only, but also to other cattle. " I do not find," says he, " that the farmers in England know the great advantage which may be derived from the use of salt, in the business of fattening cattle ; whereas in America we think it in a manner absolutely necessary; and accordingly give it to almost every kind of cattle : and those with parted hoofs are particularly fond of it.

" There cannot be a greater instance of this fondness, than the wild cattle resorting to the *salt-licks*, where they are chiefly killed, we give this name to the salt springs, which in various places, issue naturally out of the ground, and form each a little rill.

"Horses are as fond of salt as black cattle; for with us if they are ever so wild, they will be much sooner brought to a handful of salt, than any kind of corn whatever.

"We also give salt to our sheep; and to this practice it is generally ascribed, that the American cattle in general, are so much more healthy than the same animals in England: *certain it is that they are subject to much fewer diseases.*

"There is one very advantageous practice we have, which I cannot enough recommend to the notice of the farmers here in England, it is mixing salt with our hay ricks when we stack it, which we call *brining*.

"Just before I left America, I had a crop of hay which was in a manner spoiled by rain, being almost rotted in the field; yet did this hay spend as well as if it had been got in ever so favourably.—When my servants were making up the stack, I had it managed in

the following manner : as soon as a bed
of hay was laid about six inches thick, I
had the whole sprinkled over with salt;
then another bed of hay was laid, which
was again sprinkled in like manner ; and
this method was followed till the hay
was stacked. When the season came for
cutting it to the cattle, I found that so
far from refusing it, they ate it with sur-
prising appetite, always preferring it before the sweetest hay that had not been
in this manner sprinkled with salt."

The fattening effects of sea marsh is
also universally known ; and sheep both
feed on it with great desire, and also
with the greatest benefit. So wonderfully nourishing is it, that Dr Johnson in
his work on quadrupeds confirms it by
saying " that sheep become fatter in the
maritime salt marshes of Italy, than on
any other kind of pasture."

To avoid prolixity, we here cut short our
remarks on this subject. We have already
detailed at sufficient length the nutritious

effects of salt and saline pastures. Happy is it for the owners of stock, where nature has provided the latter, and fortunate also for them, where art has thought of devising the former. No desponding accounts have been transmitted to us of the prevalence of disease amongst their flocks: there the Braxy and Rot are never heard of, and are so entirely unknown perhaps, as never to have been designated by any appellation. Those marshes, which, had the herbs produced in them been uncompounded of salt, might have been favourable toward the excitement of such diseases, not only form a rampart against their admission, but are over and above the readiest food for increasing the condition of the stock by which they are pastured. Nay, even in those places where their pasture bears every resemblance to those in this country, and where they are in every respect similarly treated, except that salt is considered an essential part of their provender, we are positively informed that they are subject to far fewer diseases than those in this country, and that the complaints which prevail here, make no fearful inroads there.

Though these facts are fully accredited and have, for many years, been known, and though our nation is able to vie with almost any other in her improvements, it is matter of astonishment that salt has not, prior to this period, been considered of inestimable advantage, to the improvement of stock, although its effects as a cure of disease might have been overlooked. It admits not of the smallest degree of doubt, that sheep in every season, whether it may be favourable or unfavourable for producing the rot, would be materially benefited by salt, as tending highly to strengthen their constitution, and to better their condition. The extravagant price of that article, indeed, of late years, has put it in a great measure beyond the power of farmers to use it in ordinary seasons, merely as a means for improving stock: though it might undoubtedly have been employed in extraordinary ones, to put a check to the destructive consequences of disease. At the present reduced rate, however, the sum that would be required to provide sheep with it, would, besides the security it affords against loss, be fully repaid, by the superiority of the returns which they would make.

But to speak more immediately to the purpose. Whatever profit may result from salt being every year regularly given to sheep, I consider that it is both *a preventative and a cure* for the rot, whatever may be the originating cause of that disease. And as in almost every case, so perhaps also in this, it is easier to prevent than to cure, if salt is to be employed at all with the view of averting the rot, it will undoubtedly be proper to administer it at an early period of the season. To those who

choose the safer method of allowing it
to their flocks on the return of every autumn, which practice I would strongly recommend, it is not necessary to enquire
whether the season have a tendency to
generate the complaint, as if it have this
tendency the effects of the salt will
never suffer it to obtain any firm hold;
and if it have not, the effects of the
salt in improving the stock become the
more obvious and profitable.

But where this plan is not thought
worthy of practice, or where it would
be attended with inconveniences to do
so, however we may succeed in curing,
I am afraid it will hardly be possible
to prevent the disease. Every change
that takes place in the state of the weather throughout the month of September
and till the middle of October, must be
carefully attended to, observing if it have
a tendency to excite the rot. If any
suspicions are entertained of its being the
consequence of the then prevalent state

of the weather, the application of the preventative should not be a moment delayed.

It will only be on those farms where sheep are easily collected, that the present measure can be taken to prevent the rot; for unless they are assembled together in considerable numbers, I can conceive no proper manner, in which they can be supplied with salt, without sustaining less or more injury. The easiest method for accomplishing this purpose, is as follows:—Let a small piece of ground, about an acre perhaps for 20 score of sheep, and more or less in proportion to their number, be set apart and inclosed with nets or in any other way the farmer may think proper, about the latter end of August or beginning of September. The ground thus appropriated, must be occasionally sown over with salt; 20 bushels, of *56lbs.* each, I should think would be perfectly sufficient for one acre. Or the half of that quantity of salt may be sown and the other half dissolved and sprinkled upon the grass after it has sprung to a considerable height.   Or it may be applied in any other manner in which the farmer may think it will more readily rise in the vegetation of the grass. When the piece of ground has been thus prepared, and saved for about three weeks, the sheep should be put on it, and kept there for 24 hours, or to such time as they have eaten up the grass; after which they may be turned out, and the grass allowed to grow for two or three weeks longer, when the same thing may be repeated as circumstances may require.

Or what would perhaps serve the purpose better; two inclosures may be made, if for the same number of sheep, each half the size of the above, and pastured alternately, each inclosure, after its being first eaten, remaining at rest for two weeks; and thus the sheep would have the salted pasture every week, which might be continued for any length of time, that the owner may deem proper. And it may be observed, that if any of these measures are adopted from year to year, the same ground should always be appropriated.

If all this be rightly done, I should be little inclined to dread the approach of rot. For should the disease begin to form before the application, or during the intervals between the application of salt, the disease being only incipient, the salt conveyed through the medium of the grass, will immediately counteract it while it exists only in embryo. Though it is evident that by these means the rot can easily be prevented, taking for granted the efficacy of salt, which is well known to be destructive of insects, I cannot state this as the result of my own experience, as no rot has been of general occurence, since the destructive year of

z

1817, when such a preventative was not then applied by me.

If the above method of administering salt, which is found to be attended with such evident advantages to the flocks in a sound state, should be deemed either impracticable, or inefficient, there is no insuperable difficulty in the way of their being supplied with it, in the same manner as obtains in Spain. When the state of the weather will admit of it, let hollow stones, placed at proper distances on the sheep walks, have a proper quantity of dry salt laid on each, and let the sheep, one parcel after another in proportion to the number of stones, be trained to lick it, and after tasting it once, they will return of their own accord as often as it may be necessary to treat them in this manner. This may be done at any time when the weather sets in dry between the beginning of July and the end of October. If it be objected that this process is attended with too much labour and expense, even to be attempted, the same objections were once made to the smearing of sheep, and to their being supplied with hay and turnips in the storms of winter. As the causes of rot have not yet been sufficiently developed, it is presumed that if the salt be administered before the beginning, or at least before the middle of September it will prove the more efficient preventative of the disease, as it is better to check it in the bud than to combat it when it has assumed a formidable appearance. In this way too they will have the full use of the whole of the salt, as no portion of it will be lost by evaporation, as will be the case when it is sown upon the pasture and moistened by the dews of the night, and as it will reach the stomach in the full possession of all its natural qualities.

I have only to regret that these means of prevention and cure, in their applica-

z 2

tion on some farms, must necessarily be
obstructed by almost insurmountable difficulties.   But supposing their impractica-
bility in some instances or even their
inefficiency where not properly or in due
time applied, I trust I have proof enough
of the possibility of curing sheep of
the rot, after the disease has really begun,
by virtue of salt.

In the year 1817, one very active shepherd in my neighbourhood who had the
charge of 200 ewes, observing them
tainted with rot, bethought himself of
trying the experiment, and conducted it
in the following way :—whenever he saw
any one or more of them shewing unequivocal symptoms of rot, he brought
them into a dry court yard or empty
house, and fed them with hay, turnips,
or a few oats ; to every two of them he
gave twice a day a handful of salt, which
he dissolved in water, and putting the so-
lution into a tea-pot, it was poured down
their throats. This was repeated for se-

veral successive days, and continued till some improvement in the condition of the sheep was discernible, after which they were turned out to the field.   If the re-appearance of symptoms did not justify their continuance with the flock, they were again conducted home and the salt was as before administered.   Few of them required more than two such courses; but a great proportion of the flock was treat-ed in this manner, and the shepherd de-livered the whole of them alive at Whit-sunday, except one ewe which had died in lambing.

Another unquestionable instance of the efficacy of salt as a cure for the rot, I can also produce.   In the above year, to which I have had occasion so often to refer, I sold a lot of 80 ewes, for a trust con-cern on a neighbouring farm, in the be-ginning of October.   They went to the lower part of the county, where they got an abundant supply of wholesome food. The purchaser disposed of 10 of them at

a fair run, to a friend of his whose farm
lay along the sea coast.   The 70 that remained were soon discovered to be infected with rot, and the only means of
restoration that were applied, was a plen-
tiful allowance of the best food.   The
event proved, however, as has but too
often happened, that in trusting to the
powerful tendency of this, the farmer
trusted to a broken reed; for during the
winter and spring more than half of them
died, and only a few of what survived
were able to nurse their lambs.   They
were killed in the following Autumn, and
were all found far gone in the rot.

The 10 that went to the sea coast
never shewed any outward symptoms of
the complaint, but on the contrary got
into excellent condition, and each of
them brought up a lamb in great order.
Had our proof extended no farther than
this, there would have been a presump-
tion that the salt spray upon the sea
coast had prevented the rot from making

any greater progress, if it really had been begun. But the result leaves not the least shadow of doubt; for when these were killed in Autumn, the ravages that the fluke worms had made on the liver, were perfectly visible, the worms themselves were dead and the parts healed up; nothing but the scars remained, shewing where they had been.

Both these examples will undoubtedly, to every unprejudiced mind, afford the most undeniable evidence of the truth of salt being a remedy for the rot. To the experience of those mentioned, I can scarcely add that of my own. I have, indeed, sometimes selected a sheep from amongst the flocks, of which I have had a pretty well grounded suspicion of its being affected by the disease, and have by salt restored it to an equality with the rest. But I have not as yet had any fair opportunity of effecting a general cure, neither do I at all long for one, though I am perfectly convinced that, did the disease again appear it would be quickly removed by the use of salt properly applied, and in sufficient quantity.

In taking measures to stop the further progress of rot, after it has evidently begun, I would recommend the method of *brining* hay, which the Americans employ. When the hay is to be stacked in Autumn for winter feeding, I would have a small stack appropriated for the purpose of being sprinkled with salt. This stack to contain as much hay as would serve the sheep for a fortnight, half feeding, which to old sheep will be about a half pound per day. Provision will of course be made according to the number of sheep. When the hay is stacked, it may be sprinkled after the manner which the letter from an American, formerly quoted, points out; and I would allow about two bushels of salt to every ton of hay, which will amount to a very small expense. This ought to be given

them in any frosty morning, should any
symptoms of the disease appear ; and
would greatly increase their value were
it given any time during the winter,
though we were quite sure of the total
absence of rot.   In every season this is
a practice which I would earnestly recommend.

If this plan prove unsuccessful, of
which I trust there will be little danger,
there remains only one other resource,
and that is, by administering dissolved
salt to those sheep that still seem to be
affected, by means of a tea-pot, or any
other convenient instrument.   One sheep
may be allowed nearly half a pound of
salt every day, and to be given at two
different times.   Or one half of this quantity may be dissolved and given in the
above manner, whilst the other half may
be sprinkled upon hay.   But if the salt
be administered in any way, it is a matter of indifference, I believe, after what
manner it is done ; provided it be con-

tinued till the sheep has made a manifest improvement in its appearance, which is presumptive evidence that the disease no longer preys upon it.

If these measures, then, are carried into execution, I may affirm with some degree of confidence, that our flocks will not henceforth be carried away by this baneful malady. A wall of defence is erected against the batteries, with which it has assailed them, which will repel its severest attacks. Its pestilential influence, which has so long haunted and ravaged the fields on which our herds delighted to graze, may now stalk abroad without striking terror and dismay into our breasts. Still it may injure, but it cannot destroy; its arrows being stript of their venom, will inflict wounds only to be healed.

Braxy

APPENDIX.

———

AFTER having treated so fully of the cause, the means of prevention and cure for the rot, it may be wondered why I have been silent upon that other very destructive disease, called Braxy or sickness; this being the next most commonly known, and most fearful malady, that prevails on the generality of stock farms in this country. As I consider it, therefore, to stand, by many degrees, nearest in importance to the rot, and as it does not fall within the body of the treatise, I have here affixed the thoughts that I have been enabled to collect concerning it.

Perhaps it may be thought, that it is at least unnecessary for me to offer any remarks upon this subject, since Mr Hogg has already treated it at considerable length. There are none who will more readily allow the value and superiority of his treatise upon this disease than I do; but as many of my readers may have never had any opportunity of examining Mr Hogg's work, and as my own experience with this disease has been pretty extensive, too much so indeed, it may not be improper to state in a cursory manner what is of weight concerning it. To those, however, who wish to have a more minute acquaintance of it, I refer them to the work alluded to.

It is not at all my design to enter into any description of those different species of Braxy on which Mr Hogg has exercised his powers, both as he has conducted his observations with acute research, and as I intend to make my limited remarks entirely of a practical nature. With that species which so frequently pervades districts such as this, and which proceeds from indigestion, I mean to be altogether confined.

The remote cause of sickness is generally considered to take its rise in the hogs eating too voraciously of soft herbage in Autumn, and after the consumption of which, they are obliged to betake themselves to harder and less easily digested food. This bears very much the semblance of truth, and appears farther true from this circumstance, that I have known hogs die of it, which had come from a coarser to a finer pasture, whilst those that had been bred and reared upon the same farm were never infected. Such as had been accustomed to less valuable pasturage, would naturally devour with greater avidity the more desirable food to which they were transferred, and would thereby bring on the disease of which we speak.

To attribute the complaint wholly to this cause, however, does not seem to be altogether free from embarrassment. For being at one period very much annoyed by it, and experiencing many very heavy losses, I was led to try the experiment, amongst the many other preventatives and cures that have been prescribed, of changing the pasture of the hogs. I have put them upon a field of very good young clover, which was, indeed, attended with the effect of giving a temporary check to the disease; but after their continuance of a few days, returned with unabated violence. This instance appears to be rather at variance with the cause above mentioned, but if fully investigated might perhaps be found to accord with it, as I never made any minute enquiries into the matter.

Hogs generally die very rapidly of this distemper, and are not unfrequently found lying dead in the morning, without any symptoms having been observed, in the preceding evening. If any means are to be taken to prevent its effects, the flocks should be occasionally looked to during the course of the night. The first symptom of the disease as mentioned by Mr Hogg, is the animal's ceasing to chew the cud. " As the distemper advances, the agony which it suffers becomes more and more visible. When it stands it brings all its four feet into the compass of a foot; and sometimes it continues to rise and lie down alternately every two or three minutes. The eyes are heavy and dull, and deeply expressive of its distress. The ears hang down, and when more narrowly inspected, the mouth and tongue are dry and parched, and the white of the eye inflamed."

There is no disease perhaps, undoubtedly there is none to which sheep are subject, for which as many preventatives and cures have been prescribed, and with more flattering promises, than the one under consideration. The powers of tar

and mustard and such nonsensical stuff
have been set forth in a wonderful light,
as being infallible cures for the sickness,
but which rather hasten than prevent the
death of the animal. Those recommended by Mr Hogg, are, first upon its being
discovered the shepherd should give the
hog a severe heat by running ; next bath-
ing it amongst warm water for eight or
ten minutes, giving it at the same time
water gruel, mixed with butter. These
are simple experiments and are at least
worthy of trial.

There is something of a different cast,
which I would recommend to the trial of
those who are troubled with the braxy,
as for many years I have happily not
been troubled with it on my own farm,
and which for aught I know, has never
been singly administered for this disease,
and that is salt. As the disease proceeds more immediately from indigestion,
it is exceedingly likely that salt if given
in sufficient quantities may dispel it, as

we know that has a laxative tendency. If a large doze of this were granted the animal as soon as it was observed to be affected, we are from the ordinary effects of salt led to conclude that the result would be favourable.

But whether salt may not, from the rapidity with which the disease increases, produce the end desired, could it be given in small quantities to the whole flock of sheep among which it begins to prevail, there is little doubt that it would be effectual in preventing it. Such sheep as pasture upon the sea coasts and have the inestimable advantage of marsh, are not at all subject to the braxy, being regularly purged by the salt which is impregnated in the marsh. And we have every reason also to suppose, that if our hogs, immediately upon the commencement of the disease, were to be daily allowed a little salt or saline pastures, there would be small loss occasioned by the braxy.

In some districts, especially in the Etterick forest, the farmers have in a great measure banished the sickness, by "pasturing the young and old of their flocks all together." This plan wherever it can be practised without much inconvenience, should always be carefully attended to, as the young reap advantage in various ways from thus living in common and sharing the experience of the old. But this is in many cases almost impracticable, at least would be attended with no profit. On such farms, as was before observed, let the pasture for one hirsel be as nearly as possible of one soil; to overlook this is a mighty error, and the surest means of making the flock unequal. The heather should also be regularly burned, and the sheep never allowed to pasture long upon soft grass.

There is one other measure only which I intend to recommend, and which I consider to be an infallible antidote against the progress of this malady. If the above attempts to prevent it, either fail or cannot be practised, the next and the last method to which recourse can be had, is to put the hogs upon turnips. If it continue to destroy them rapidly, they should be immediately inclosed in a small portion of a field of turnips; for if their bounds are too wide, the disease will for some time continue, until they begin to eat the body of the turnip. This I have invariably found gives a settling stroke to the disease. It may indeed, commence its ravages so early, as to occasion the loss of a few turnips; but this will be comparatively small to what might otherwise be sustained.

Besides these, I know of no other antidotes against this disease. If it cannot be put a stop to by any of these means, it will baffle all our efforts. But if these are carefully applied, I conceive that I have sufficient grounds to warrant me in stating, that the braxy, like the rot, will no longer deprive us of the best of our flocks, and that it will henceforth be known by little else but its name.

Finis.

www.ingramcontent.com/pod-product-compliance
Lightning Source LLC
Chambersburg PA
CBHW081719220526
45468CB00008B/1899